Partners in STEM Press
Portland Oregon

Copyright 2014
by Jeannie Ruiz

Published by Partners in STEM Press
7401 N Burlington Ave
Portland, Oregon 97203

All rights reserved. No part of this publication may be reproduced, stored in a retrieval system, or transmitted in any form for any reason, recording or otherwise, without the prior written permission of Partners in STEM Press.

ISBN 978-1-942357-30-8

Photo permissions and authors, illustrators, designers, photographers available for download at the URL.

http://www.ten80elementary.com

STEMWORKS EXEMPLARY PROGRAM STATUS

Ten80 Education and its programs are included in the STEMWorks Database by Change the Equation, a non-partisan, CEO-led initiative to connect and align efforts to improve STEM learning. Ten80 Education programs met the highest standards for excellence and exemplary STEM project based programming through rigorous examination by WestEd, an independent nonprofit research organization.

RECOMMENDED BY REAL TEACHERS

We are thoroughly enjoying our STEAM lessons. The program is excellent! My students have loved making the Rhombis and getting to build with shapes and tape. It has been a big hit! Thanks for everything!
Alicia Stenard
Mater Christi in Albany School District

"We need more projects like this that carry on for a longer number of weeks. My teachers used scope and sequence suggestions, and the program ran after school for a semester. We could have worked on this a lot longer."
Karen James
Druid Hills Elementary in Charlotte, NC

"I'm excited for our students to get their hands on real STEAM projects. There isn't much out there for Kindergarten and First Grade."
Sandy Mettler
Fort Worth Schools in Dallas, TX

FIRST GRADE MODULE 1: USING THIS STORYBOOK

JUMP-START INVESTIGATIONS WITH STORIES
Each project and STEMvestigation has an associated story. The story does not offer the recipe for success in each STEMvestigation. Rather, the literacy component is a jump-start to discussion on the topic. Images are limited to encourage kids to offer their own vision. The goal of this book is to encourage the "bedtime story" effect. They will interact with plenty of books that walk them through with pictures. Let students tell you what they think Quad looks like or what a Saturday Farmer's Market might sell.

THESE STORIES ARE WRITTEN ABOVE GRADE LEVEL
These stories are intentionally written above the level of a first grader's reading ability. Children will not build mature syntax or extensive vocabularies if they are limited to six-word sentences. Building the ability to "see" a story can only happen when the child hears the story and begins to imagine.

READ AND RE-READ
Kids don't get bored with repetition in television or books. Read the same book several days in a row. Hold a discussion each time about the images, or let kids draw their own versions. Use this book just like you'd use any compilation. You are the teacher. You know when the storybook read-aloud component is appropriate for your students.

S.T.E.A.M. IN THE CONTEXT OF SCIENCE CLASSROOMS
Recurring themes are pervasive in sciences, mathematics, and technology. These ideas transcend disciplinary boundaries and include patterns, cycles, systems, models, and change and constancy.

KINDERGARTEN
- Students observe and describe the natural world using their five senses.
- Doing science as inquiry develops and enriches understanding of scientific concepts and processes.
- Students develop vocabulary while investigating common objects, earth materials, and organisms.
- Counting, place value, numeracy, and measurement are given useful context within the scope challenges.
- Students should begin to recognize math as the language of science.

FIRST GRADE
- Students observe and describe the natural world using their five senses.
- Students develop/enrich abilities to understand the world in the context of scientific concepts and processes.
- Students develop vocabulary investigating properties of common objects, earth materials, and organisms.
- Students are encouraged to use their budding mathematical skills as useful tools in the pursuit of solutions.
- As designers, students will activate these skills in inquiry-focused projects.

SECOND GRADE
- Students observe and investigate to learn about the natural world and reveal patterns, changes, and cycles.
- Students should understand that certain types of questions can be answered by using observation and investigations and that the information gathered in these may change as new observations are made.
- As students participate in inquiry-driven and project-based investigation, they begin to understand the integration of mathematics in the context of scientific investigation and the design process.

FIRST GRADE MODULE 1: USING THIS STORYBOOK

JUMPSTART INVESTIGATIONS WITH STORIES

Each project and STEMvestigation has an associated story. The story does not offer the recipe for success in each STEMvestigation. Rather, the literacy component is a jump-start to discussion on the topic. Images are limited to encourage kids to offer their own vision. The goal of this book is to encourage the "bedtime story" effect. They will interact with plenty of books that walk them through with pictures. Let students tell you what they think Quad looks like or what a Saturday Farmer's Market might sell.

THESE STORIES ARE WRITTEN ABOVE GRADE LEVEL

These stories are intentionally written above the level of a first grader's reading ability. Children will not build mature syntax or extensive vocabularies if they are limited to six-word sentences. Building the ability to "see" a story can only happen when the child hears the story and begins to imagine.

READ AND RE-READ

Kids don't get bored with repetition in television or books. Read the same book several days in a row. Hold a discussion each time about the images, or let kids draw their own versions. Use this book just like you'd use any compilation. You are the teacher. You know when the storybook read-aloud component is appropriate for your students.

FIRST GRADE MODULE 1: TABLE OF CONTENTS

UNIT	RHOMBI-TALE	PAGE
WORLD GOES ROUND		8
SCIENCE	Luna and Sol's Big Adventure	10
TECHNOLOGY	Rover, Rover, Come on Over	12
ENGINEERING	Wheels on the Ground	14
ARTS	Quilting in Circles	16
MATHEMATICS	Radius and Pi	18
SQUARE ONE		20
SCIENCE	Square Foot Harvest	22
TECHNOLOGY	P Is for Parallel	24
ENGINEERING	Shopping for Shapes	26
ARTS	Stories in the Square	28
MATHEMATICS	Paper Forests	30
WE LOVE TRIANGLES		32
SCIENCE	Billy Goat's Bluff	34
TECHNOLOGY	City of Three Angles	36
ENGINEERING	Traveling Triangle	38
ARTS	Fox's Forest	40
MATHEMATICS	Tri, Tri, Again	42
TAKE FIVE		44
SCIENCE	We're Hunting for a Pentagon	46
TECHNOLOGY	Tool Time	48
ENGINEERING	Creative Design	50
ARTS	Design Time	52
MATHEMATICS	The Pentagon, USA	54
TINY HOUSE ON WHEELS		56
GLOSSARY		58

STEAMstart

MODULE 1: TWO DIMENSIONAL SHAPES

 1
 2
 3
 4

EXTENSION

WORLD GOES ROUND
UNIT 1

Once upon a time, there was a kid named Rhombi. Rhombi liked shapes and saw them everywhere.

One sunny afternoon, Rhombi and her uncle headed down the street to get some ice cream. Rhombi's Uncle Galileo was an astronomer. He looked for stars, planets and exciting actions in the universe.

Rhombi suddenly turned to her uncle and asked, "Why don't we see the sun all night?"

He smiled and looked up into the night sky. He scratched his short hair, and answered, "That has to do with things going around in circles. It has to do with the way the moon revolves too.."

Rhombi's uncle began to spin.

"There is part of your answer," said Uncle Galileo. "I'm the Sun at the center of the Solar System. I spin in place. All the planets spin in place. All the moons spin in place."

Rhombi began to spin. Her uncle said, "Now here is the fun part. You spin and circle my chair." Rhombi was spinning and walking around his chair. Uncle Galileo started spinning in place at the center of Rhombi's circle. The two got so dizzy they had to stop.

"The moon spins and revolves around Earth. Earth is always rotating. Part of every day, we are turned away from the Sun," said her uncle. "At the same time, Earth is revolving in a big loop around the Sun. It's like wheels inside wheels inside wheels."

Rhombi thought of all the things she knew that could spin. There were pinwheels and wind-up toys and spinning tops. Uncle Galileo loved all of them.

His birthday was coming, and Rhombi had an idea. She looked out from the pages of her world and asked, "Designers, would you help me create a rotating gift for my uncle?"

Luna and Sol's Big Adventure
Explore circles.

Every day was an adventure for Luna. She loved to travel and never, ever sat still. Her favorite dance in the world was a twirling rotation. Like a ballerina on a jewelry box, Luna could spin and spin and spin.

Luna's best friend was Sol. Sol lived in the same neighborhood as Luna, but Sol was a lot bigger and older than Luna. Where Luna was cool and calm, Sol was fiery and grand. Where Luna liked to stay near home, Sol traveled in great revolutions around the Milky Way.

Even though they were different, Luna and Sol stayed in touch no matter how much Sol traveled. Sol kept an eye on Luna, warming her with his fiery light. Luna had no light of her own. The side of Luna that faced away from Sol was always cold and dark. But the side that faced him was warmed by Sol's light. Each month - well, closer to 29 and a half days - Luna tuned around one full turn...

Luna's whole life was a dance. Big Sister Earth danced around Sol while Luna danced ahead and sometimes behind her, always smiling when Sol poured his light and warmth on her face.

Luna and Sol's Big Adventure
Explore circles.

Each month, as they moved through the First Quarter of the dance, Luna twirled between Big Sister Earth and Sol,

Big Sister Earth could not see Luna. When that happened, Luna and Sol smiled at one another and Luna twirled faster so Big Sister Earth could start to see her again.

Sol was so big that he cast light on Luna, even if Big Sister Earth was in the way.

Sometimes, Big Sister Earth managed to dance between them so that Luna was in her shadow and Sol could not see Luna.

That did not happen very often.

Luna would giggle and peek out at Sol after a few minutes. Little by little she moved back into Sol's view and continued her dance.

Sol, Luna and Big Sister Earth twirled in their places, rotating in easy circles as they changed places through their dance each month - well, closer to 29 and a half days. Luna's whole life is a dance.

Rover, Rover, Come on Over
Explore circles.

Two rovers sit still on the moon's surface. The dust on their roofs had not moved in many years. No wind had stirred the little flags attached to the rover antennae. There was no atmosphere on the moon to create weather or wind.

That's Heavy, Man
The LRVs or Lunar Rover Vehicles were very delicate and could never be driven on Earth. They were specially designed for the moon's gravity. Earth's gravity is much higher, making everything on Earth feel heavier than those same objects would feel on Earth.

Apollo Missions 15, 16, 17
The two rovers had been used many times in three lunar visits. Apollo 15, Apollo 16 and Apollo 17 astronauts used the same rovers, parking the vehicles between missions. The Lunar Rover Modules never drove farther than 4.72 miles from the rocket's landing place.

Hammer Time
On the Apollo 17 mission, the hammer in Commander Gene Cernan's space suit pocket accidentally broke one rover's fender. If astronauts hadn't repaired the fender, moon dust would have clogged delicate machinery and equipment.

4.72 miles

According to Space.com, the rovers measured about 10 feet long (3 meters), 6 feet wide (nearly 2 meters) and almost 4 feet (over 1 meter) high.

Rover, Rover, Come on Over
Explore circles.

Duct Tape Designs

On the moon, there are no mechanics to fix a fender, so the astronauts had to use materials they had on hand. The Commander and his team created a new fender using a laminated map and duct tape. You could see the actual EVA* map fender on display at the National Air and Space Museum.

Men on the Moon

The rovers have sat still and silent for more than 40 years. The last visit to the moon launched on December 7, 1972. At the end of that mission, the rovers were present as astronauts placed a permanent plaque on the moon.

"A17-plaque" by NASA - http://images.jsc.nasa.gov/lores/S72-55417.jpghttp://science.ksc.nasa.gov/mirrors/images/images/pao/AS17/10075902.jpg. Licensed under Public domain via Wikimedia Commons - HTTP://commons.wikimedia.org/wiki/File:A17-plaque.JPG#mediaviewer/File:A17-plaque.JPG PUBLIC DOMAIN

Wheels on the Ground
Explore circles.

Wheels on the Ground
Explore circles.

Radius and Pi were friends. They looked the same. They even dressed the same. Radius and Pi were very different people. Radius loved things that were fast. She did not stop to think about what could happen. Pi was a planner. Pi made lists and always made decisions after lots of thinking. Radius and Pi made a good team.

On Saturday, the two always pulled a cart of organic fresh vegetables and fruits to the local Farmer's Market. The twins grew their own food to sell. They set up a table beside all the other farmers and gardeners. Radius and Pi never missed a market.

This weekend Radius and Pi had a problem. Rain had made muddy puddles on the market path. The cart's skinny tires had sunk 2 inches in the mud. The two were stuck.

Pi sat down to think. She made notes on a pad and listed all the ways they could solve the this problem. She came up with trucks and sleds.

Radius dove into his project. He liked the idea of a sled. He found a big piece of plastic and piled all the fruits and veggies on the sled. Radius yelled, "Ta Da!" He pointed to his solution and dragged the sled a few feet. It worked but was heavy.

Pi decided to try the truck idea. He found four tires from a toy in the shed. Radius helped Pi prop up the cart. They used a screwdriver and wrench to remove the cart's wheels. Radius held the cart steady as Pi connected each wheel to the cart's axle. The two lowered the cart to the ground.

Radius gently pulled the handle. He tugged a bit harder. Finally, the cart started moving. The new, wider wheels worked pretty well! Their cart sank only about 1 inch into the mud. The two moved all the produce to the wagon. The wagon rolled along the path all the way to market.

Radius and Pi sold all the food and had a great day. Best of all, they had worked as a team to solve a tricky problem. Radius and Pi always made a good team.

Quilting in Circles
Explore circles.

Quilting in Circles
Explore circles.

Radius and Pi were visiting Aunt Bee. She had promised to teach the two how to make a quilt. Aunt Bee's quilts were works of art. Their aunt loved shapes. Today she was helping the two finish a project. It was full of bold circles sewn by hand.

Aunt Bee had already made the basic quilt. Its bottom layer was squares of old t-shirt fabric. The center layer was fluffy cotton like pillow filling. The top layer was bright fabric, and two were sewing on this layer. Aunt Bee gave each two a thimble to keep fingers safe from sharp needles.

Radius and Pi had already used a template to cut circles in lots of sizes. Radius cut circles that measured 1 inch from center, 2 inches from center and 3 inches from center. They had 22 circles in different colors to sew.

Pi drew a pattern of circles that made lots of dots on the fabric. Some of the circles touched edge to edge. Other circles sat one inside another in a quilting pattern called an echo. Each circle was sewn by hand. They hand-sewed their circles one by one.

The planet was rotating away from the Sun when Radius and Pi finished their project. The Sun sank out of sight, and they could see the moon in the sky. The two put down their needles. They shook out the beautiful quilt. Circles overlapped and danced on one side. Bright t-shirt squares filled the back. The quilt would always remind them of a great day with Aunt Bee.

Radius and Pi
Explore circles.

Radius and Pi
Explore circles.

Radius and Pi rolled along the path toward Saturday market. Suddenly, the cart stopped. Radius circled the cart looking for a problem. He leaned down to take a look at the wheel and axle. The wheel had wobbled loose and cracked. Radius looked around for some way to fix the cart. The two needed a new wheel. Radius and Pi began looking for round objects that could fit the cart. The two found a roller skate wheel and a skateboard wheel in a pile by the curb. Both wheels would be too small. The length from the center hole to the outer part was just too short. Their cart might roll, but it would tilt and spill the apples. They found some balls, but none had a hole in the center. No hole meant no axle. How would the ball stay on without an axle down the middle?

Radius tried to hold up the no-wheel corner while Pi steered. The two entrepreneurs really wanted to reach the market in time to sell their fruit. They moved too slowly without a wheel. Pi and Radius looked at the bicycles in a shop window. They looked at each other with wide grins and carefully propped the cart on a step. Four minutes later, Radius emerged with a bicycle wheel minus the tube. It was old and rusted but not bent. The shop owner gave them the wheel. The two gave her 10 oranges. The wheel was just the right length from center to rim.

Radius and Pi put the wheel on their cart. They zoomed to the market. When the market gate opened, customers were happy to see the twins. Radius and Pi sold all the yummy ripe apples. The two were proud that they'd solved an engineering problem all on their own.

SQUARE ONE UNIT 2

SQUARE ONE
Explore squares.

Once upon a time, there was Rhombi. Rhombi loved shapes. She had a telescope to look at the big round moon. Her innovative friends had helped Rhombi design a cart that pulled her telescope to the beach for stargazing.

Rhombi wondered what else she could find with her telescope. She turned its lens to the Earth. Rhombi saw lots of shapes. She saw circles and all kinds of quadrilaterals. In the yard and around the neighborhood, Rhombi saw a lot of objects with 4 edges and 4 vertices.

Rhombi knew that not all quadrilaterals were squares. Some 4 sided shapes that looked like squares were really rectangles or diamonds or kites. I wonder if the roof shingles are really squares? They could be some other kind of quadrilateral.

What had her teacher told the class about squares?

"A square has four equal edges with four corners shaped like the letter L," the teacher had said.

All the shingles looked like the same length, but she would need a ruler to double check. Rhombi noticed a few missing shingles. Looking at the playhouse through her telescope, Rhombi knew that she should fix the roof before the weather changed. Rain could leak through the gaps between shingles and ruin her playhouse.

Rhombi needed a tool that would help her make sure the shingles were really square. Then she needed a way to make new shingles that were also square.

This was a problem that needed teamwork. Rhombi looked out from the pages of her world and asked, "Builders, can you design a tool to make sure my shingles are square?"

Square Foot Harvest
Explore garden harvests in science.

Square Foot Harvest
Explore garden harvests in science.

Leaves crunched underfoot as Rhombi headed toward the garden. She kicked a dozen colors into the air just to watch them float back to the ground. autumn was her favorite part of the year.

Squirrels chattered as they collected acorns from the ground under oak trees more than 70 feet tall. The trees towered above Rhombi like a giant swaying ceiling. She knew these trees must be at least 50 years old if they were dropping so many acorns.

Rhombi and her friends had planted vegetable seeds in the Community Garden in August. Over the past 2 months, the seeds had germinated, pushing out of their hard seed covers. Seedlings began to push through the soil toward the sunlight. Rhombi had spent many hours thinning the rows so each seedling would have enough sun and room to grow. When she saw leaves, Rhombi knew that the radish bulb and roots were reaching down into the soil below. Finally, after almost 40 days, the radishes were ready to eat.

Opening the gate, Rhombi nearly tripped over the huge pumpkins in the first part of the garden. Pumpkins needed lots of room to grow. Each seed had to be planted 5 feet apart. Growing 10 pumpkins took almost 50 feet of garden space. Someone had planted mini-pumpkins that climbed a small trellis at the end of the pumpkin row.

Carrots and radishes and lettuce waited for Rhombi. She pulled 20 carrots and put them into her basket. Next she layered 24 radishes. Finally, Rhombi added 4 heads of lettuce. She picked up her heavy basket and headed home through the crunchy leaves.

She washed and chopped carrots and radishes. She tore lettuce leaves, and added them to a big bowl. When her friends sat down to dinner, Rhombi would proudly serve her delicious autumn harvest salad!

P is for Parallel
Explore shapes in technology.

P is for Parallel
Explore shapes in technology.

A is for Angle, where two lines meet up...
B is the Base, a 3D shape's bottom or top.
C is for Coordinate, where x and y are.
D is for Decagon, like the drawing of a star.
E is for Ellipse that looks like an oval.
F is for Fractal, every part like the whole.
G is for Google, 1 with 100 zero's.
H is for Hexagon, the honeycomb's hero.
I is for Interior including all the stuff inside.
J is for Junction where things touch or collide.
K is a Kite with four triangles glued together.
L is for Line. Does it go on forever?
M is for Meter, a unit of measure.
N is for Nonagon, a nine sided treasure.
O is for Octagon. Octa means 8 sides.
P is for Parallel, two lines that never collide.
Q is a Quadrilateral, 4 sides and 4 corners.
R is for Rotation, to turn around a shape's center.
S is for Spheres that are perfectly round.
T is the Triangle that keeps a bridge from falling down.
U is for Undecagon, a 10-sided shape with no order. V
V is for Vertex, hanging out at the corner.
W is for Width, a measure from side to side.
X is for X-Axis, a horizontal line.
Y is for Y-Axis, up and down on a graph.
Z is for Zero, no dinner for giraffe?

Shopping for Shapes
Explore shapes in engineering.

Shopping for Shapes
Explore shapes in engineering.

Poly and her friend Trap were shopping for groceries. The two were meeting up with Rhombi that afternoon for chess club. It was Poly's turn to bring snacks. She had ten dollars to spend.

Poly pushed a cart up and down the aisles looking for something her friends might like to eat. She dropped a pack of gum in the cart, but it fell through the squares that covered the cart. She placed the gum on the rectangular face of a box of cookies.

Poly wanted to find cheese, crackers, fruit, and cookies. So far, she'd found cookies. They were the round kind with chocolate chips. Trap held up a bag of $3.00 star fruits. They looked like spheres or balls now. When cut, the star fruits would look like little green star shapes on a plate. He also had fat purple grapes that cost $1.00 per bag.

The list included cheese, crackers, fruit, and cookies. So far, Poly and Trap had cookies and fruit. The two friends headed toward the dairy section to look for cheese. Wow, there were a lot of cheese shapes. Which one should they choose? There were cylinders of string cheese. Poly pointed to brick shaped rectangular prisms of cheddar. Trap spied a bag of already cut cubes that would be easy to pile on a tray. The $2.00 cheese cubes were added to the cart.

The last item on their snack list was crackers. The cracker aisle was filled with boxes of all shapes and sizes. Oval buttery crackers sat next to 2 inch square saltines and little ½ inch round soup crackers. There were whole wheat rectangles woven like blankets and cheesy 1 inch squares. Some were even shaped like goldfish. Trap picked the $2.50 woven rectangles. They tasted good with cheese.

Their cart was filled with delicious shapes. Star fruits, grapes like spheres, cubes of cheese, and square crackers filled the cart. The total cost was $8.50. They had one dollar and 50 cents left. Poly and Trap took their bags and headed to the chess club. Snack time shaped up to be a yummy break.

Stories in the Square
Explore squares and patterns in art.

Stories in the Square
Explore square and patterns in art

Quad had a problem. His class was making a quilt for the autumn fundraiser. Each student was responsible for creating a pattern and making one square called a block. All the blocks would be sewn together to make one big quilt. He had no idea what to make.

His floor was littered with crumpled papers. Quad could not choose a pattern. He had drawn and colored and traced until his fingers hurt, but none of the drawings seemed right for his piece of the class quilt. What was Quad going to do? His brain needed a jump-start.

Quad put down his pencil and headed out to chess practice. He would think more about the quilt block later. At the recreation center, Quad sat down at his usual table. Rhombi sat across the chess board, placing pawns in a row. Behind the 8 pawns, she placed 1 king, 1 queen, 2 rooks, 2 bishops, and 2 knights on the board. The kids took turns moving pieces across the 64 black and white squares. Rhombi almost always played red, while Quad played the white pieces.

As they played, Quad got an idea. He hurried through the match, said a quick goodbye to Rhombi, and hurried home. Grabbing a new piece of paper, Quad folded a square paper in half 3 times. Opening the paper, Quad counted 8 columns. He folded the paper in half 3 times in the other direction. Now he had 64 equal squares.

Quad colored every other square black. He added 8 red pawns, 1 red king, 1 red queen, 2 red rooks, 2 red bishops, and 2 red knights to the black and white board. Quad used his pattern to cut the fabric for his quilt block. He carefully sewed all the pieces together and held the finished quilt block to take a look. It was a really cool pattern.

When the class quilt was pieced together, Quad proudly added his chess board. That week, at chess practice, Quad asked Rhombi for a favor. She agreed with a laugh as Quad set up the Red chess pieces and got ready to play.

Paper Forests
Explore origami in mathematics.

Paper Forests
Explore origami in mathematics.

Rhombi's Grandpa folded beautiful paper art. When Rhombi was little, Grandpa could make animals and flowers and boats and castles appear from his hands using nothing but squares of thin paper. Grandpa taught her to fold paper boxes to hold paper jewels. He taught Rhombi to make beautiful princesses with swords that rescued dragons from wicked wizards. Rhombi's favorite project was a mobile hung with 40 folded animals. Grandpa called it his forest in the sky.

Grandpa no longer folded paper, though. His hands were old and gnarled. The fingers couldn't hold delicate paper anymore. Now, instead of the country, Grandpa was living in a city home with lots of other grandmas and grandpas. They all played checkers and had holiday parties. Grandpa seemed happy, but sometimes he talked to Rhombi about the old days. They laughed about a funny fast squirrel that sneaked in through the window and raced around the bedroom before bursting out the front door.

Grandpa remembered pushing Rhombi on the tire swing. He talked about animals that visited from nearby woods. Grandpa especially missed the red fox that used to peer in through the bedroom windows on cold snowy nights. Sometimes they found it sleeping in his shed. Grandpa's new home was in the city. No animals peeked in city windows.

Rhombi planned to visit Grandpa for his birthday on Saturday. Sitting at her desk, Rhombi thought about a gift. She knew he missed his forest and fields, but Rhombi could not bring any live animals into Grandpa's new home. Looking at all the origami papers on her table, Rhombi decided that she could bring some animals to Grandpa after all. She picked up a stack of papers and started to fold.

Saturday was the big day. Rhombi eased her gift through the front door of Grandpa's building. Other grandmas and grandpas and aunties and uncles smiled as she made her way to the dining room. All Grandpa's friends were there with cake and balloons. Everyone oohed and ahhed as Rhombi entered. Grandpa turned to look, and his hands flew to his mouth. He smiled. He grinned. He laughed out loud!

Hanging from Rhombi's hands was a mobile full of forest animals. His squirrel was there in its tree. There were blue birds and rabbits. But, best of all, hanging bright and sassy at the top of Grandpa's new forest in the sky was a beautiful bright red fox. Happy birthday, Grandpa!

WE LOVE TRIANGLES
UNIT 3

WE LOVE TRIANGLES
Explore triangles.

Once upon a time there was Rhombi. Rhombi loved shapes. Circles were cool (there were circles in the word "cool"). Her innovative helpers had designed a tool to find squares. That had been a fun project.

Rhombi sat in the park. She looked around the playground and played her own game of "I SPY" with shapes.

"I spy with my little eye a square tile."
"I spy with my little eye 80 circles."

80 circles were linked to make a chain that held the swing. She had counted them one day at recess.

The swing seats were squares. She knew because each seat had 4 equal sides and 4 corners shaped like the capital letter L.

For the first time, Rhombi saw that the swing set had a big triangle at each end of the top bar. Rhombi watched her friend Quad swing. She looked at the triangles holding up the swing set. The bases of the triangles at each end never moved.

She also saw that the bridge from slide to hanging bars was made of many triangles. The triangles sat tight against each other underneath the bridge's wooden planks.

Triangles were everywhere.

After recess, Rhombi's class had center time. Rhombi went to the hamster cage to clean the cedar. She would refill the pet's water, too.

The hamster sat on its nest. The habitat was a cube. It had a red base. The top and sides were clear plastic so kids could see the hamster. The poor rodent looked bored. Rhombi wished she could give the pet recess. Kids used recess to recharge their brains. The hamster looked like it could use a brain break.

Rhombi smiled and said, "Well, if I can't take you to recess. I can bring recess to you!"

Rhombi looked out the window. She saw the playground bridge. She saw the swings. Then Rhombi saw the slides.

"That's it," she thought. "I will make a slide for our class hamster."

Rhombi knew she would need some help. She looked out from the pages of her world and asked, "Friends, will you help me design and make a slide with a sturdy triangle base for our hamster? He weighs about as much as a beanbag."

Billy Goat's Bluff
Explore Bridges in Science

LEFT COLUMN
MAN MADE
THESE BRIDGES

RIGHT COLUMN
NATURE MADE THESE BRIDGES!
a. beaver b. fallen log c. eroded rock

Billy Goat's Bluff
Explore Bridges in Science.

Once upon a time there were three little goats. They were hungry and wanted to eat clover at the top of the triangle hill. To reach the clover, each goat had to cross a rickety rackety bridge. Youngest goat began to cross the rectangle bridge. He dashed across the 12 feet of bridge as it swayed back and forth under his weight. Rickety rackety rickety rackety went the bridge. The beaver under the bridge shouted at them to go away. "You are too loud, and you're making my bridge go rickety rackety."

Youngest Goat answered, "I hope the square-bottom bridge will hold. My brother is on the way, and he is much bigger than I."

After 30 minutes, Middle Goat began to cross the bridge. Rickety rackety rickety rackety went the bridge. The beaver under the bridge lumbered out and shouted at them to go away. "You are even louder than your brother, and you're making my bridge go rickety rackety."

Middle Goat answered, "I hope the square-bottom bridge will hold. My brother is on the way, and he is much bigger than I."

Along the path came Biggest Goat. He took one look at the bridge and shook his head slowly. His triangular coal black beard waggled back and forth. The biggest goat put one hoof on the bridge. Rickety rackety rickety rackety went the bridge. The angry beaver underneath shouted at them to go away! The goat peered under the bridge and saw that the bridge was held up by a single square. He looked across the stream at his brothers. If he couldn't fix the bridge, the younger kids would be stuck on the hillside. Biggest Goat thought and thought. What was the strongest shape he could remember?

Youngest Goat saw a seed and bleated, "What about a circle?"

Biggest Goat shook his head. The bridge would roll away. Middle Goat saw a flower petal and bleated, "What about an oval?"

Biggest Goat shook his head. The bridge would rock back and forth like a rocking horse. He needed a very powerful shape to keep the bridge straight and strong. Finally, Biggest Goat looked up at the mountain that held the delicious clover. What shape is a mountain? It's a triangle. Worn by weather and wind, the mountain stands strong for thousands of years. Biggest Goat asked the beaver to help him with the bridge. The beaver grumbled but made lots of triangles to hold up the bridge. Finally, Biggest Goat began to cross the bridge. The happy beaver waddled back under his very strong bridge. He didn't even notice when the three brothers headed back across the bridge at sunset.

City of Three Angles
Explore maps in technology.

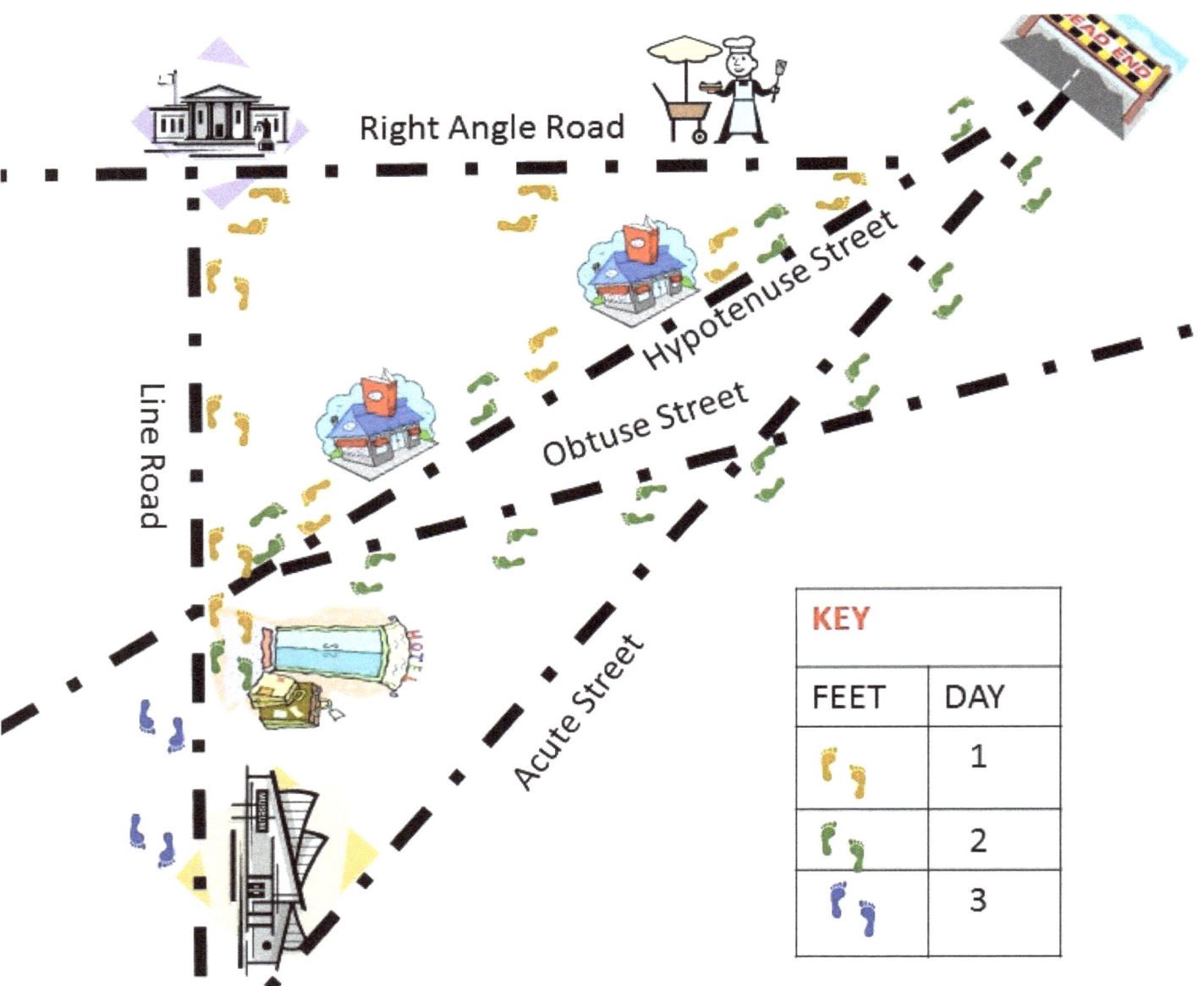

City of Three Angles
Explore maps in technology.

Every street in Three Angles was named after the shape of its corner. That made walking through town easy if you knew your shapes. Quad wanted to find Polygon Museum. He was new to the city, but he was pretty sure he could get around. Quad didn't bother taking a map.

Quad walked North along Line Street to the town square. He turned right at the square and walked East along Right Angle Road. Quad bought a cool dog collar at the Pet Palace. At the farthest end of Right Angle Road, Quad took a sharp right turn and headed South along Hypotenuse Street. He found two great bookstores along the way. Finally, Quad ended up right back where he'd started! He had not found the museum. He decided to try again on Friday.

The next morning, Quad looked for the Museum again. This time, he started on Hypotenuse Street. He headed North on Hypotenuse until the street came to a dead-end. Quad decided to turn right onto Acute Avenue. He was sure to find the museum at the end of Acute Avenue. Quad reached the corner and took a slight right onto Obtuse Place. He walked and walked until he saw the sign for his hotel at the end of Obtuse Place just before Hypotenuse.

Good grief! The sun was setting. He'd have to find the museum on Saturday. This time, he'd save time and use a map.

On Saturday morning, Quad stood outside his hotel. The young tourist looked at his map. He touched the name of the Polygon Museum on the map. Slowly, Quad looked left. The museum was beside his hotel. Had Quad turned left out of the hotel, he would have seen the Museum right away. Next time, Quad would definitely use a map.

Traveling Triangle
Explore triangles in engineering.

Traveling Triangle
Explore triangles in engineering.

Triangle, Triangle,
where have you been?
I've soared in the sky
blown East and West by the wind.

Triangle, Triangle,
what have you seen?
I've watched the city
from the top of a Ferris wheel.

Triangle, Triangle,
what have you done?
I've held the wood steady
for builders in the hot summer sun.

Triangle, Triangle,
where have you gone?
I've biked along city streets
in the cool quiet dawn.

Triangle, Triangle,
what should we play?
Let's climb on the jungle gym
at the end of the day.

Fox's Forest
Explore environments in art.

Fox's Forest
Explore environments in art.

Fox trotted happily on the warm summer path. Sun beamed down through leaves to make moving patterns on the worn dirt track. Fox's head swung from side to side looking for yummy treats. He also scanned the woods for anything that might want to eat him!

Bushy tail wagging, Fox stopped to nose some flowers. Soon these green leaves would part to reveal little red berries. Delicious and sweet, Fox loved these berries in the straw. Fox caught another smell in the straw. It was the ripe odor of rotting log. Logs were good. They often held tasty crunchy bugs.

Fox lifted his little pointed face to the wind. Where was the smelly log? He faced West toward the setting sun. Sniff, sniff. No logs there. He faced East toward the far edge of the woods. No log smell rose from that direction. Fox turned North to sniff. There! The stinky log was that way!

Nose to the ground, Fox left the path and headed into the green woods. Shadows played across Fox's rust-colored body. Sniff, sniff. He followed the air molecules that blew in the breeze from that stinky log. The molecules were pulled in through Fox's nose and mouth. His brain made sense of the smells, and Fox turned a little to the West. Finally, behind a tall rock, Fox found a log so rotten it made his mouth water. Using tiny paws and claws, Fox pulled at the wet bark. Beetles and mealworms tried to scatter. Fox caught them and crunched his way through an afternoon snack. Delicious!

Pointy little face covered in soil and bits of rotten tree, Fox turned back toward the path. South and a little East brought Fox right back to strawberry leaves beside the warm summer path. It wasn't quite so warm now that the Earth was turning away from the Sun. Shadows crept closer as Fox found his way back to the den. With a full belly, Fox curled up with his tail over his nose and fell asleep in the safety of his dark, dry hole.

Tri, Tri Again
Explore triangles in mathematics.

Tri, Tri Again
Explore triangles in mathematics.

Poly really needed to fix her bird feeder's roof. The feeder hung from a branch outside her window. All kinds of feathered friends relied on the feeder for their winter meals. Sadly, there were a bunch of furry friends that also wanted to eat at the hanging seed source. Long fluffy tails knocked the feeder to the ground over and over again. Poly was afraid that the feeder would break in one of its squirrelly falls.

Poly had tried lots of different ways to trick the squirrels. She wanted to keep them away from the feeder, but the mammals wanted all the free seeds. Squirrels had nuts of their own, but the bird feeder was such an easy food source. Poly had designed a round tipsy cover that should have made the squirrels wobble and fall to the ground. They learned how to climb around it in less than a week. Poly had turned a mesh scarf into a wrap that squirrels couldn't get their hands through. They'd just untied and dropped the rectangular scarf to the ground. Poly had even covered the bird feeder hanger with drops of slippery oil. The greedy beasts just slid down it like a fireman's pole. They even seemed to have fun!

Poly had thought she was out of ideas. Round plates hadn't worked. Rectangular scarves hadn't worked. Spheres of oil hadn't worked. That afternoon, on her class trip to the city museum, Poly had seen a tall pointed roof. The roof covered a bell, and the guide told her class that the roof kept pigeons from landing. Bingo! If it kept pigeons away, the roof should keep squirrels away.

Poly cut out 3 triangles of wood. She covered them with waterproof material used for raincoats and connected all three triangles at their long edges. She formed a tall pointed object and attached it to the bird feeder. Her brand new triangular roof was nearly 3 feet tall and very steep. She hung the modified feeder and sat down to wait.

Birds flew in and out to get their seed meals. Finally, two hungry little squirrels came to sit on the branch above the feeder. They tilted little heads to one side and the other side looking confused. Finally, one brave little friend leaped to the roof. She slid right past the seed and landed on the ground. That was one grumpy squirrel. Her rodent friends each tried the feeder, but all failed. Poly had learned her lesson.

If at first you don't succeed with circles and rectangles, tri, tri again.

TAKE FIVE
UNIT 4

TAKE FIVE
Explore pentagons.

Rhombi, Quad, and Poly put on hiking boots. They fill bottles of water. All three friends wear long pants and long sleeved shirts. Quad hands each friend some sunscreen, and they all put on a hat.

The friends will look for deer, snakes, raccoons, birds, and turtles. The kids open a map of Pentagon Park with 5 edges and 5 vertices. It's a pentagon.

"Hey, the Nature Center trails are laid out in a pentagon," Poly shouts. "That's like the shape on my soccer ball."

Quad laughs but shushes Poly. "We will never see any animals if you shout."

The kids take a trail leading to the deer meadow. With any luck -and quiet voices- they will see some animals in the woods. They hike for half an hour.

"Look at the deer," whispers Poly.

Peeking from the edge of the woods are two deer. Many deer have used this place to cross the path. They are perfectly still and silent.

Suddenly, there is a commotion. Mountain bikers speed down the trail toward the deer. Poly shouts at the deer to move faster. She shouts at the bikers to stop!

The bikers have permits and are being safe, but an accident could happen any second! The deer are startled by Poly's shouting, and they begin to move.

Mountain bikes skid to a messy stop. Gravel and dirt fly into the air just as the last deer bounds away into the trees.

"If this is a place the deer walk a lot," says one biker, "there should be a sign to let us know." She is breathing hard.

Rhombi, Quad, and Poly finish their hike. At the end of the trail, they sit down to rest. After talking about the deer crossing, Quad says to his friends, "Let's fix this problem. We can make some signs to warn people on the trail that deer might be nearby."

The others agree and ask Rhombi if her friends might help with the project.

Rhombi looks out from the pages of her world and asks, "Designers, will you create deer crossing signs like the pentagon-shaped school crossing signs outside your school?"

We're Hunting for a Pentagon
Explore pentagons in science.

We're Hunting for a Pentagon
Explore pentagons in science.

We're going on a 5 hunt. (We're going on a 5 hunt.)
We're looking for a pentagon. (We're looking for a pentagon.)
We're gonna catch a big one. (We're gonna catch a big one.)
I'm not afraid. (I'm not afraid.)
Are you? (No way!)

Here comes the crossing sign.
Better wait for the crossing guard.
Walk briskly and pump your arms.

REFRAIN

We're passing a tall house. It sure is wide and has 5 edges. Let's walk around it.

REFRAIN

We're walking through the soccer field. Uh-oh. Here comes the soccer ball. We better duck under it…

REFRAIN

We're running through the garden now. It sure is grassy. I see the morning glories and okra. Jump over the rows. Oooohhh. We're all muddy. Shake it off….

REFRAIN

I see a big building. It has 5 sides. It's The Pentagon.
Oh no, what time is it?
Recess starts in 5 minutes. We're going to miss it.
Better get home fast!
Oh no, mud! Shake it off.
Jump the okra and morning glories.
Yikes, duck the soccer ball.
Walk quickly around the tall house.

STOP! The crossing guard says to wait. Okay, we can cross now. Walk briskly and pump those arms. Hurry! Hurry!
Ahhhhh, we made it!

Tool Time
Explore tools in technology.

Tool Time
Explore tools in technology.

Hilda and Poly had a job to do. Grandma wanted the garage organized. She liked everything in its place. The two friends had taken out the trash. They had bagged cardboard, newspaper, glass, and cans for recycling. They had even swept and dusted.

The tool bench was still a big mess. Tools were scattered on the table. Tools were dumped in boxes under the bench. Tools were hung in all the wrong pegs on the wall.

Hilda and Poly started by sorting the tools used to move items. They hung hammers and mallets on the peg board over jars of nails in lots of sizes. Star-shaped Phillips head and flathead screwdrivers used to tighten or loosen things went into bins by size. #1, #2, and #3 were matched to their same sized screws.

Rhombi put grabbing, twisting, and pulling tools on the pegboard. Pliers were hung from smallest to largest. Wrenches were hung over jars of nuts and bolts.

Rhombi and Quad did not touch the cutting tools. Saws, hacksaws, pincers and wire cutters were left for Grandma to handle later.

Carpenter's levels, rulers, angles, squares, and other measuring tools were hung neatly to the right of the pegboard wall behind the bench. A caliper used to measure very tiny items was placed in its box. Some tools were delicate.

When Hilda and Poly finished the job, they proudly called Grandma to see her shiny garage. The excited old woman clapped her hands and hugged both girls. When Grandma offered them money for doing the job, Hilda refused to take the money.

"No, Grandma," said Hilda. "I'm happy to help!"

Creative Design
Explore pentagons in engineering.

How Many Pentagons Can You Count In This Set Of Pictures?

Creative Design
Explore pentagons in engineering.

Pentagons have 5 edges, 5 edges, 5 edges.
Pentagons have 5 edges,
regular or not.

Regular pentagons have 5 edges
all the same, all the same, all the same.
Regular pentagons have 5 edges
all the same length.

Irregular pentagons have 5 edges
not the same, not the same, not the same.
Irregular pentagons have 5 edges
NOT the same size.

ALL pentagons have 5 angles, 5 angles, 5 angles.
All pentagons have 5 angles
regular or not.

The angles of all pentagons
add up, add up, add up.
The angles of all pentagons
add up to 540 degrees.

5 Edges
5 Angles
500 + 40 Degrees
PENTAGONS are fun!

Design Time
Explore polygons in art.

http://alturl.com/g7sgj

Design Time
Explore polygons in art.

Young man, you amaze me. --*Albert Einstein, physicist (1938)*

Buckminster Fuller said, "Whenever I draw a circle, I immediately want to step out of it." Buckminster Fuller was an architect and an engineer born in 1895. He was also a designer and an inventor. Fuller wrote 28 books on subjects in math and science. He had 47 special degrees from different universities.

Fuller was someone that cared about making the world better. He was also a humanitarian, someone that wants to improve life for other people. Each year the Buckminster Fuller Institute offers a $100,000 prize to "scientists, students, designers, architects, activists, entrepreneurs, artists and planners from all over the world using innovative solutions to solve some of humanity's most pressing problems." From a floating doctor's office to a new plan for food production, Fuller's prize helps people with great ideas afford to put their plans in action.

This inventor was interested in using the planet's resources more efficiently. He invented a geodesic dome made of polygon shapes. The dome uses fewer materials and can be made bigger or smaller to meet the needs of many different types of uses. Gardens, homes, and museums are shaped like geodesic domes.

When Buckminster Fuller first imagined the special dome, he thought of it as humans working like a crew to survive on our planet, "spaceship earth." Epcot Center in Florida created a domed building inspired by Fuller's idea.

Domes like these are made of triangles. As more triangles are added, the shape gets stronger. No other 3-dimensional shape can be so big without extra supports inside the building. The larger it is, the stronger the dome becomes. These domes are so simple that an entire house can be built in a single day. They are so strong that geodesic domes can even survive hurricane force winds!

Can you see the triangles in the playground dome?
Can you see diamonds in the playground dome?
Can you see hexagons in the playground dome?

The Pentagon, USA

HOLDS A LOT OF CARS

🚗 = 10,000

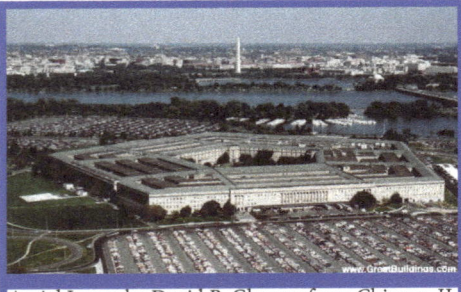
Aerial Image by David B. Gleason from Chicago, IL (The Pentagon) via Wikimedia Commons

COUNT on the PENTAGON

1 — formal dining room
2 — cafeterias
3 — days to design the building
4 — thousand 2 hundred clocks
5 — edges and vertices in a pentagon
6 — snack bars
7 — minute walk corner to corner
8 — decades since building opened
9 — hundred 21 feet per outer edge
10 — thousand visitors a year

5 Stories Tall +2 below ground

5 Armed Services housed on site

5 Acre Central Plaza "Ground Zero"

Executive Order 8802 let The Pentagon open all its 284 bathrooms to all employees, black or white. For a time, The Pentagon was the only non-segregated building in Virginia.

NUMBER OF MONTHS FROM GROUND BREAKING TO OPENING CEREMONY											
	JAN	FEB	MAR	APR	MAY	JUNE	JULY	AUG	SEPT	NOV	DEC
1941									■	■	■
1942	■	■	■	■	■	■	■	■	■	■	■
1943	■										

Reference Material
January 15, 2013 9 Things You May Not Know About the Pentagon By Barbara Maranzani via History.com
Vogel, Steve. The Pentagon: A History. New York: Random House, 2007. ISBN 1-400-06303-5 (pp. 199-200, 216-217)
History of the Pentagon — U.S. DOD info, archived at Archive.org.
Pentagon Web Site — By the U.S. Department of Defense.
Http://en.wikipedia.org/wiki/The_Pentagon and Wikimedia Commons

Graphic Information
Explore info-graphics in mathematics.

Info-graphic is short for information graphic. An info-graphic uses pictures and words to make lots of information and numbers easy to grasp. Look at this info-graphic showing information about The Pentagon. You could read the paragraphs that follow, or you can use an info-graphic to show the same information.

The Pentagon was designed in just 3 days and built far faster than most government buildings. It was also built to hold a lot of cars. Parking was planned for 10,000 cars. 16 months passed from the first day of construction until employees parked all those cars in the lots. The construction trucks, earth movers, and builders showed up for work on September 11, 1941. The Pentagon was completed in January of 1943. The building opened almost 8 decades ago.

The building serves a lot of needs. Very special visitors are invited to parties in the formal dining room. Most armed services workers eat in one of the 2 cafeterias. 5 different armed services use the building. 10,000 or more visitors use any of the 6 snack bars.

Finding a bathroom in The Pentagon is easy. When the building was planned in 1941, Virginia state law still said that black people and white people had to use different bathrooms. The pentagon was built with 284 bathrooms. By order of the President, none of those bathrooms were ever separated. Employees and visitors of all backgrounds used any of the Pentagon's public bathrooms. For a time, The Pentagon was the only non-segregated building in the state of Virginia.

Getting from place to place in the huge building takes time. Each outer edge measures 921 feet. There are 5 stories full of offices and a 5 acre central area called "ground zero." The total square footage of the The Pentagon is more than 5 million square feet. Your school probably measures less than 50,000 square feet. Walking from one corner to another corner of The Pentagon takes nearly 7 minutes. How many seconds pass if you walk from one corner of your building to another corner of your building?

Which format seems easiest to use? Which format gives information quickest? Which format offers more of a story? Do you like the paragraphs or graphic style best?

TINY HOUSE ON WHEELS
Extension for Module 2

Rhombi and her friends are planning a camping trip for Friday night. Quad, Poly and Rhombi have made lists of things they will need. Rhombi looks over the list one more time.

They have sleeping bags, water, and snacks. They each have one flash light and extra batteries. Each kid has a warm sweatshirt for the cold part of the night. The kids have almost all their gear ready.

Quad, Rhombi, and Poly are missing two important things.

Rhombi looks at the list and sits down to think. She has been

	RHOMBI	QUAD	POLY
sleeping bags	x	x	x
apples and bananas	x		x
water bottles	x	x	x
tortilla chips and salsa	x	x	x
extra sweatshirts	x		x
flashlight	x	x	x
tent or wagon	?	?	?
a way to move stuff	?	?	?

reading a book about the early settlers. They moved across the country in covered wagons. Each wagon held what a family would need to start a whole new life. The wagon was also the family's home on the long trip West.

Rhombi had an idea. She was sure her friends were innovative enough to design a sort of covered wagon. She, Poly, and Quad could use the wagon to carry their supplies. They could also sleep in the wagon if the weather turned rainy.

Rhombi looked out from the pages of her world and asked, "Innovators, will you design and build a model of the wagon we could use for camping? We will need to move three people and all of our supplies. Your scale model will need to carry at least 6 ounces of weight for a length of 3 feet."

FIRST GRADE MODULE 1: 2D SHAPES
VOCABULARY

Most vocabulary is defined within each story's context. The definitions on these pages address vocabulary outside the grade level and undefined within the story itself.

astronomer
a person who is skilled in astronomy or who makes observations of celestial phenomena

autumn
also know as fall, this season lasts from the September equinox to the December solstice

cylinder
a solid object with two identical flat ends that are circular or elliptical and one curved side.
It has the same cross-section from one end to the other.

designer
a person who plans the form, look, or workings of something before its being made

engineer
a person who has scientific training and who designs and builds complicated products, machines, systems, or structures : a person who specializes in a branch of engineering

FIRST GRADE MODULE 1: 2D SHAPES
VOCABULARY

pi

the 16th letter of the Greek alphabet and the symbol π denoting the ratio of the circumference of a circle to its diameter

radius

length of a straight line from the center of a circle or sphere to any point on the outer edge

revolution

the action of moving around something in a path that is similar to a circle

rhombus

a shape with four sides of equal length and with four angles that are not always right angles

rotation

a complete turn around a central point

solution

something that solves a problem

telescope

arrangement of lenses or mirrors or both that gathers light and allows direct observation of distant objects.

FIRST GRADE MODULE 2: TABLE OF CONTENTS

UNIT	RHOMBI-TALE	PAGE
RHOMBI'S PLAYOUSE		60
SCIENCE	Wind on the Windows	64
TECHNOLOGY	Accidental Inventions	66
ENGINEERING	The Tiny House	68
ARTS	Cube World	70
MATHEMATICS	Do You Know the Quadrilaterals	72
ROOF ROOF		74
SCIENCE	Water Is as Water Does	76
TECHNOLOGY	The Case of the Disappearing Sand	78
ENGINEERING	Circle, Triangle, Square	80
ARTS	I Love Homes	82
MATHEMATICS	The Camel and the Pyramid	84
PET FINDS A HOME		86
SCIENCE	Davis Finds a Purrl / Purrl Finds a Home	88
TECHNOLOGY	Rhombi Lived in a Zoo	90
ENGINEERING	It All Adds Up	92
ARTS	M.C. Escher's World	94
MATHEMATICS	How Big Is a Guinea Pig	96
HAPPY BIRTHDAY, PET		98
SCIENCE	That's Bananas	100
TECHNOLOGY	Waxing Colorful	102
ENGINEERING	Strength in Numbers	104
ARTS	Square in the Middle	106
MATHEMATICS	Coming up a Cloud	108
RHOMBI'S CANDIES		110
GLOSSARY		111

MODULE 2: THREE DIMENSIONAL SHAPES

EXTENSION

RHOMBI'S PLAYHOUSE
Explore cubes.

Once upon a time, there was Rhombi. Rhombi loved shapes and found them everywhere. She especially loved the squares that made up each face of a cube.

Rhombi had a new tool set, and she wanted to build a playhouse. Rhombi chose a shape for her house that looked like blocks and sugar cubes.

Rhombi knew that she should build a sturdy house to withstand the fall winds, but the she also wanted to go apple picking. Rhombi hurried to finish the playhouse so she could run and play.

That night, the weather changed. The air grew colder and blew wildly around the corners of Rhombi's home. Rhombi walked to the edge of the field to watch the rainstorm.

Returning home, she jumped in surprise as one sharp gust of wind pushed at her house. Another gust of wind pulled at the house. The walls shuddered. The house shivered. So did Rhombi.

She stared at all the crispy leaves as they tumbled from the trees. Then, Rhombi stared in amazement as the walls of her house also tumbled to the ground.

Looking at the walls piled on the cold ground, Rhombi decided she needed some help. She looked out beyond the pages of her world and asked, "Builders, can you help me build a new and stronger house?"

Wind on the Windows
Explore weather in science.

WORDS FOR WIND

CHINOOK
westerly wind off the eastern side of the Rocky Mountains

SANTA ANA
easterly towards Southern California

SCIROCCO
southerly from North Africa to southern Europe

MISTRAL
northwesterly from central France to Mediterranean

MARIN
southeasterly from Mediterranean to France

BORA
northeasterly from eastern Europe to Italy

GREGALE
northeasterly from Greece

ETESIAN
northwesterly from Greece

LIBECCIO
southwesterly towards Italy

Wind on the Windows
Explore weather in science.

My dog is scared. She is shaking and whining under the bed. I sit on the rug and rub her ears. It does not seem to help. Daisy is a big floppy German Shepherd, and she is not scared of anything. Well, she is scared of one thing. Daisy hates the sound that wind makes as it hits the corner of the house.

I try to tell her it is just the air moving. I speak calmly, and I stay very quiet and still. All these things help Daisy feel safe. I describe the sounds I hear. There are four tall pine trees outside the window. The wind through their needles sounds like a quartet of violins. Air pushes past the pine needles like a bow on violin strings. A willow tree by the creek trails its long branches into the water. Air lifts the branches and whistles through leaves like a dozen flutes singing.

Daisy is shaking a little less. I think she likes my story. Her head rests on my leg, and big brown eyes look up at me through tan lashes. Her tail thumps a bit. I keep talking.

Sound starts as vibration. I thump Daisy's nametag to make it jingle. Sound hits your ears like a pendulum swinging. Little tiny hairs in your ears start to vibrate. All those branches and leaves shaking outside are vibrating the air, see? Daisy's eyes roll to the window, but she does not lift her head. Faster vibration means high pitched sound. Listen.

We close our eyes and listen to the wind. All that moving air pounds the window panes, making each piece of glass rattle like a drum. Some of the wind slips through the cracks around the window, and we hear deep mellow sounds. "Those are drums and cellos in our symphony," I tell Daisy. She is not impressed.

Windy flutes, violins, cellos, and drums play. With the moon high above and the faded rug below us, Daisy and I drift off to sleep. Some wild and blustery version of Bach's Cello Suite No.1-Prelude flies through the night. Daisy calms down as we listen to the symphony of wind.

Accidental Inventions
Explore new ideas in technology.

1.0 POPSICLES

11 year old Frank Epperson wanted to make soda pop at home. In 1905, the popular drink could only be bought at stores and restaurants. Frank used his porch as a testing lab. He mixed powder and water to find the right taste and texture. Frank left the ingredients in a cup where he'd been working. Temperatures dropped overnight. Frank's drink froze with the stir stick stuck in its center.

Soda Pop + Icicle = Popsicle

2.0 WAFFLE CONES

The 1904 World's Fair is the birthplace of today's popular ice cream cone. Ice cream in dishes had been around for years. A dessert seller at the Fair was doing so much business that it quickly ran out of bowls and plates. A waffle maker at the next booth was not doing much business at all. The two business owners worked together to roll up the Persian waffles and fill them with ice cream.

Ice Cream + Waffles = Ice Cream Cone

3.0 CHOCOLATE CHIP COOKIES

The Toll House Inn's owner was named Ruth Wakefield. Ruth wanted to make some of her delicious chocolate cookies for people staying at the inn. Ruth realized that she was out of baker's chocolate. The inn keeper broke the regular sweet chocolate bar candy into little chips for the cookie dough. She expected the bits to melt into chocolate cookies. They did not. Instead, the chips stayed whole.

Dough + Chips = Chocolate Chip Cookies

4.0 STICKY NOTES

In 1968, Inventors Spencer Silver and Art Fry were researchers at 3M Laboratories. Spencer created a "low tack" sticky substance. It was unique in that it could be removed without hurting a surface. Unfortunately, no one had a use for the stuff. It went into storage. Many years later, Art Fry needed a way to keep papers in his choir's song book. He suggested a use for Spencer's sticky invention!

Low Tack Sticky + Paper = "Post-Its©"

5.0 SLINKY SPRING TOYS

Naval Engineer, Richard Jones, dropped one of the tension springs for a battleship project on the ground. It bounced from place to place all over the room. He spent about 2 years finding just the right materials and coil to turn his spring into a toy. Betty, his wife, named the new toy. "Slinky." is a Swedish word that means sleek or sinuous. In 1945, the first batch of 400 toys sold out in 90 minutes.

Coil Spring + Hard Work = Slinky ©Toys

6.0 ?

What items do you use that might be combined to create a new invention?

Could you find a new use for something that you already own?

What would you invent to solve a problem or make the world a better place?

The Tiny House
Explore structures in engineering.

The Tiny House
Explore structures in engineering.

My Grandpa and I are building a tiny house. Grandpa says his tiny house will use less electricity and water than the house he and Grandma used to share. I am helping. Most of Grandpa's tiny house will be built far away and delivered to him in big flat boxes. We will put all the different pieces together to make a whole house.

There is a lot to do before the boxes get here. First we prepare the site. The land must be "graded." All the rocks and roots are taken away. The land is made as flat as possible. We drive a little road grader back and forth, then use a laser to level the ground. Next comes the foundation.

Grandpa forms and pour concrete cubes called pier pads. We have to wait a couple of days while the concrete sets. Two 4ft x 8ft beams rest side-by-side on the concrete. This holds the floor underneath the walls and house. I use a wrench to thread nuts onto bolts. Grandpa follows behind to tighten each one. He lays the drainage pipe and wiring through floor tunnels before adding the last of the sub-floor.

Grandpa will use a special composting toilet. All waste runs into a system of barrels under the tiny house to become fertilizer for his garden. The water will come from city plumbing, but gray water (waste water from sinks) will drain into the garden also.

Grandpa uses a generator to power tools for framing and roofing. Grandpa's walls were pre-made. Aunt Elizabeth comes over to prop up the walls and attach frames. We "plumb" the walls to be sure they stand up straight from the floor. I hold a "level" tool so that its bubbles are perfectly centered. We make sure windows and doors are level and square. We wrap it all in sealing tape to keep out water and mold.

Next, we'll add the roof. Grandpa uses a band saw to cut the rafters that rise from walls to roof. The rafters are spaced 4 feet apart. Insulation is added to walls and roof, then layers of roofing material top off the tiny house.

The house is done! Grandpa and I paint the tiny house and add shutters. We hang a sign that says, "Welcome to Grandpa's Tiny House in the Woods."

Cube World
Explore cubism in art.

Cube World
Explore cubism in art.

In the early 20th century, artists began to look at the world as geometry in motion. Cubism began in France in 1907. Picasso and Braque started painting the world made up of cubes, spheres, cylinders, polygons, cones, and other geometric shapes. Cubists wanted to show all the sides of an object in the same picture. The paintings looked a lot like the artist had cut the image and glued its pieces back together.

Cubist paintings show objects from more than one angle at once. Pablo Picasso and Georges Braque were first to add bright colors to cubist art. They believed that painters should not just repaint the world exactly as it exists. Instead, they wanted to show every part of the whole subject.

Re-imagine everything as 2D and 3D shapes, and watch the world change before your eyes.

Do You Know the Quadrilaterals
Explore quadrilaterals in mathematics.

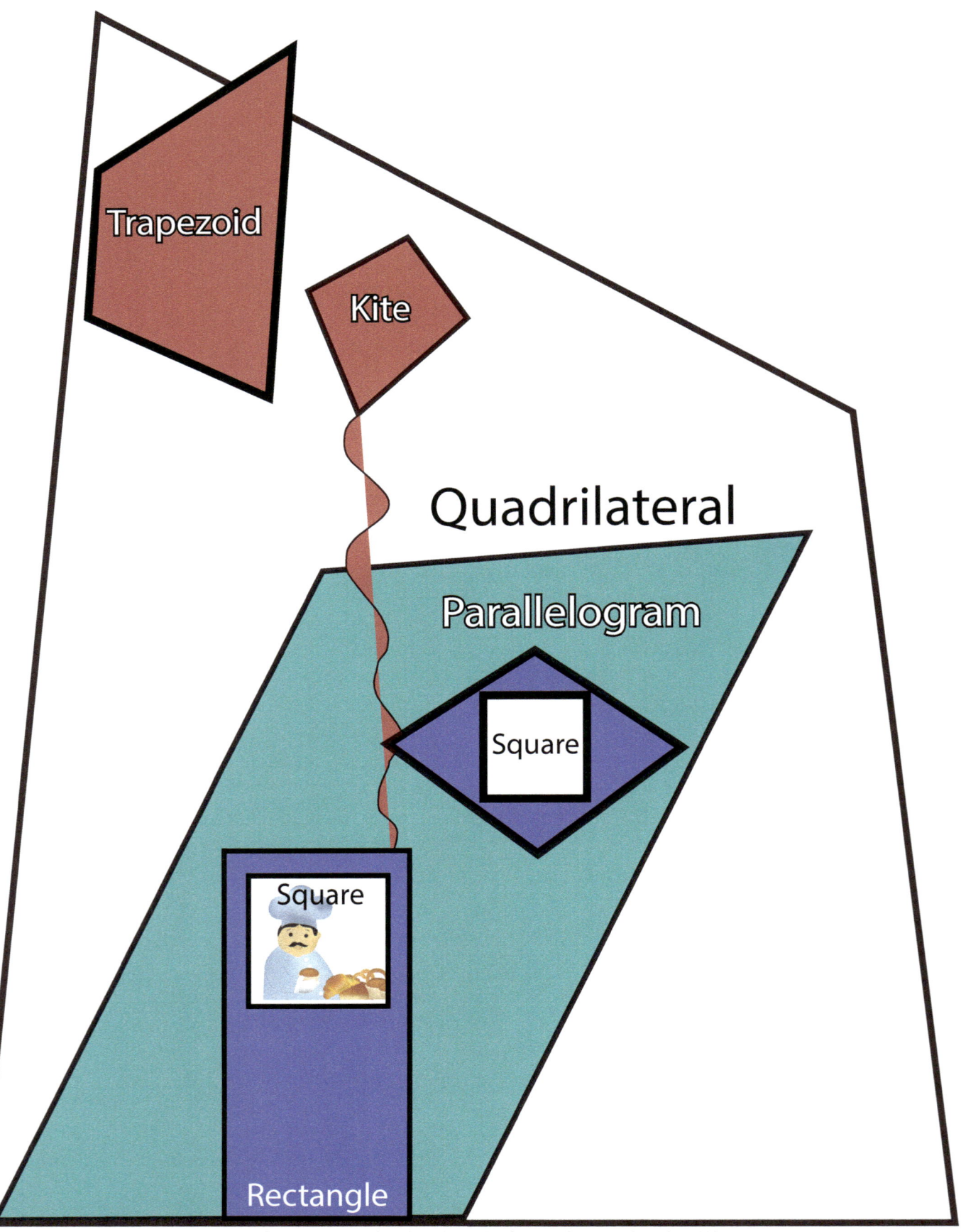

Do You Know the Quadrilaterals
Explore quadrilaterals in mathematics.

Oh do you know the quadrilateral, quadrilateral, quadrilateral?
Do you know the quadrilateral? A closed shape with 4 edges.
Oh, yes, I know the quadrilateral, quadrilateral, quadrilateral.
Oh, yes, I know the quadrilateral. 4 edges and 4 angles.

Now, do you know the trapezoid, the trapezoid, the trapezoid?
Do you know the trapezoid? It's a quadrilateral.
Oh, yes, I know the trapezoid, the trapezoid, the trapezoid.
Oh, yes I know the trapezoid. It has a pair of parallels. (a pair of LLs)

Now, do you know the parallelogram, parallelogram, parallelogram?
Do you know the parallelogram? It's a quadrilateral.
Oh, yes, I know the parallelogram, parallelogram, parallelogram.
Oh, yes, I know the parallelogram. It has TWO pairs of parallels.

Now, do you know the rectangle, the rectangle, the rectangle?
Do you know the rectangle. It's a quadrilateral.
Oh, yes, I know the rectangle, the rectangle, the rectangle.
Oh, yes, I know the rectangle… (deep breath)
It has two sets of parallel edges where opposites edges and opposite angles are the same.

Now, do you know how to find a square, find a square, find a square?
Do you know how to find a square. It's a quadrilateral.
Oh, yes, we know how to find a square, find a square, find a square.
Oh, yes we know how to find a square. Look for all edges and all angles the same.

Now we know our quadrilaterals, quadrilaterals, quadrilaterals.
Now we know our quadrilaterals.

Trapezoid, Parallelogram, Rectangle, Square!

ROOF ROOF
Explore pyramids.

Once upon a time there was Rhombi. She had a strong playhouse shaped like a cube. Autumn winds had not been able to push it down. Now that it was winter, she loved to hang out with a book and a warm blanket.

While the playhouse was snug and strong, the roof might need a redesign. Water had caused problems. Rain had made pools of water that sat on the roof. Snow and ice had frozen and thawed to make more puddles.

Drip. Drip. Drip. Freezing water and thawing ice had caused a hole in the roof! She worried that her books might be ruined. Rhombi imagined a waterfall in her house.

Rhombi's dad used a ladder to check out the playhouse roof. He looked over the flat roof full of puddles.

He said, "You have a real hole up there. I think you need to redesign the roof."

What kind of roof would she need? Rhombi stood outside with her umbrella and rain boots. She looked at her wet bird/squirrel feeder's steep roof. If she had a roof like that, water could not pool and make puddles.

She looked out beyond the pages of her world and asked, "Builders, can you help me build a waterproof roof for my playhouse?"

Water Is As Water Does
Explore the states of water in science.

Water Is As Water Does
Explore the states of water in science.

LIQUID
Water that we drink - or spill - is called liquid. Water changes names as it changes temperature.

ICE
Water that gets very cold turns solid. Solid water is called ice.

VAPOR
When water gets warm enough, it seems to disappear but is still in the air. Water we cannot see in the air is called vapor.

WATER, WATER EVERYWHERE
How many of these phrases do you know?

Wet your whistle...
Like oil and water...
Blood is thicker than water...
Mad as a wet hen...

CELSIUS
ice   0°
vapor   100°

FAHRENHEIT
ice   32°
vapor   212°

The Case of the Disappearing Sand
Explore the data collection in technology.

Read the charts with data collected by Rhombi and Pi.
What story does the data tell?
What happened to the sand sculpture?

DAY 1
THE PYRAMID BASE IS 3 FEET SQUARE
4 FEET HIGH
SUNNY
WIND 5 MILES PER HOUR

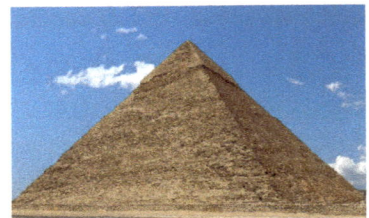

DAY 2
BASE IS ALMOST 4 FEET SQUARE
3 FEET HIGH
SUNNY
WIND 10 MILES PER HOUR

DAY 3
BASE IS 2 FEET ON THE LONGEST SIDE BUT NOT ALL SIDES ARE EVEN.
2 FEET HIGH ON ONE SIDE
ALMOST 3 FEET HIGH ON THE OTHER SIDE
(NO LONGER SYMETRICAL)
RAIN LAST NIGHT
WIND 25 MILES PER HOUR

DAY 4
THE BASE IS NOT SQUARE
1 FOOT HIGH ON ONE SIDE
ONE SIDE HAS LOST A LOT OF SAND
RAIN
WIND GUSTING TO 4 MILES PER HOUR

DAY 5
MOSTLY FLAT AND NOT MUCH SAND LEFT

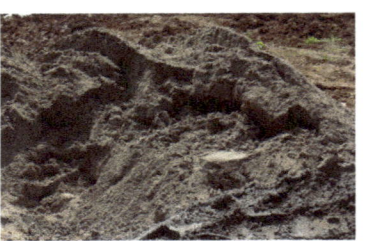

The Case of the Disappearing Sand
Explore data collection in technology.

The school bell rang. Kids tumbled out of the building. "What are you doing this weekend," Rhombi asked her friend Pi. Pi didn't have any plans. Neither did Rhombi. Maybe they would go for a swim or meet at the park?

Rhombi and Pi saw a bright orange flier stuck to the school bulletin board.
"Sand Sculpture Contest Saturday at 9AM"

The two friends looked at the sign. They looked at each other. They knew exactly what they'd be doing this weekend.

Rhombi and Pi hauled buckets, spades, small shovels, and lots of cardboard boxes to use as sand molds. The team got name tags and were assigned a 5ft by 5ft area of sand. They sketched out a quick design. Rhombi used clay to mold a model of the sculpture. When both friends were happy with the design, they started to build. They dug. They piled. They formed and pressed. Finally, the sand started to look like a pyramid. Small camels milled around at the base, and sand tents held tiny people. About 4PM, Rhombi and Pi stepped back to look at their finished sculpture. It looked great!

Rhombi and Pi watched as judges made notes at each team's area. There were some amazing designs: seagulls flew over bridges; tsunamis crashed on beaches; castles held dragons, and a couple of robots held hands. The judges called all the contestants over to announce winners. Rhombi and Pi got bright green participation ribbons. They cheered as 1st, 2nd, and 3rd place winners were given medals. Then everyone headed home, tired and sandy, but proud.

Rhombi and Pi passed the beach each day on the way to school. They noticed something interesting. The sculptures were changing shape. Each day, their pyramid design changed just a bit. The two decided to make a timeline that showed the changes every day for a week. Each day, they made a fast sketch of the sculpture and jotted down notes to keep a good record. They also tracked rain with a measuring cup system and wind with an anemometer. The weather detectives could solve this case. What was making the sand disappear?

Circle, Triangle, Square
Explore rooftops in engineering.

What is the shape of your rooftop?

Circle, Triangle, Square
Explore rooftops in engineering.

Circle, Triangle, Square

How many rooftops touch the air?

Round like the top of a castle turret.

Round like the top of a tower.

Round like the clock that's hand strikes each hour.

Slanted like the ridge of a Swiss chalet.

Angled like the top of a tepee.

Pointed like a spire in the city.

Flat like a hospital's helicopter pad.

Flat like the tar on a high rise.

Flat like the planetarium, mirrors pointed at the sky.

Circle, Triangle, Square

How many rooftops touched the air?

I Love Homes
Explore houses in art.

I Love Homes
Explore houses in art.

I love homes.
Tall homes… Small homes…
Round homes and square,
Homes in big buildings with hundreds of stairs.

I love homes.
Blue homes… Yellow homes…
Red homes and white,
Homes with black shutters to close out the night.

I love homes.
Prairie homes… Hobbit holes…
Towers and castles,
Yurts in the grassland and tents in the desert.

I love homes.
Stilt homes… Ranch homes…
Igloos and trailers,
Maybe I'll live on a boat like a sailor.

I love homes.
Cottage homes… Apartment homes…
Cabins or condos,
Tiny houses or duplexes lined up in neat rows.

I love everyone's home.

What kind of home do you love?

The Camel and the Pyramid
Explore patterns in mathematics.

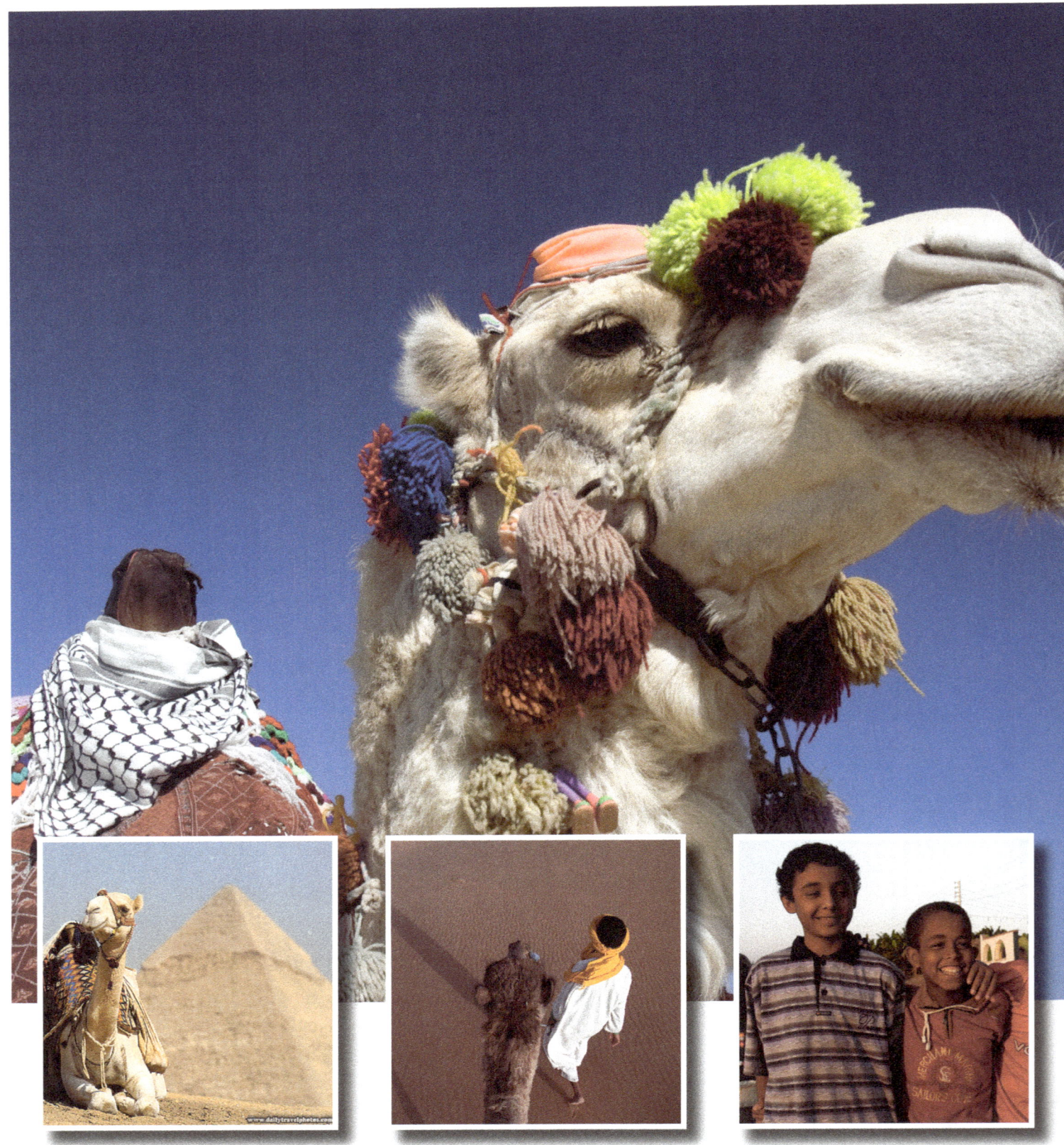

The Camel and the Pyramid
Explore patterns in mathematics.

There is a camel that lives in the far away sand of the Egyptian desert. His tan fur is stiff and rough. It has to keep out the tiny sand crystals as they crash through air during windstorms. He can close his nostrils and a third eyelid to keep out the terrible sharp sand. He has a hump that stores fat. The camel can live for days without food as his body uses the hump's fat to stay alive. Camels are tough.

The camel's caretaker is 14 years old. The young camel's guide wears a Kathye, a head wrap that does the same job as his camel's rough hide and closed nostrils. The scarf wound around the boy's head keeps sand and sun away from the boy's face. Wind blows lightly at 9 or 10 miles per hour through most days, but windstorms from the desert can be harsh. The wind may whip around the pyramids at 150 miles per hour. This kind of wind is dangerous to boys, camels, and pyramids.

The mahout knows that the pyramids will last a long time, though. As he guides tourists from all over world, he tells the history of these pyramids. He tells the families and backpacking teenagers that The Great Pyramid was made of horizontal rows of bricks. About 2,500,000 bricks were stacked like blocks. Each brick weighed around 2.5 tons. That means each brick weighed more than a car or a full-grown male giraffe. Tourists, happy with his stories gave him tips called baksheesh. The boy earns a lot of baksheesh to help care for his camel and help feed his family.

On this day in March, the winds are blowing. There are no tourists to hear his tales or ride his camel. The young mahout and the 800 pound camel head home to strong stables. They need to escape the sand and wind. As they plod away toward Cairo, the mahout looks over his shoulder at The Great Pyramid. It would not blow away in the storms, even though little bits of its faces were eroded by the constant wind. Thousands of years would pass before the wind and water slowly brought the pyramids to the closer ground.

The camel lowered his third eyelid, closed his nostrils, and headed home toward fresh water and good food and a dry place away from the itching, scouring winds.

PET FINDS A HOME

Once upon a time there was Rhombi. She had a sturdy playhouse shaped like a cube. A pyramid roof protected her playhouse from the sun, wind, and rain.

Rhombi took a walk. She saw all kinds of pets with their people. There were puppies. There was a bunny. She even saw a guy walking with a lizard on his shoulder.

After seeing all those animals, Rhombi was excited about getting a pet. She visited the animal shelter. Pet lived in Rhombi's playhouse for several months. Rhombi's mom was allergic to cats, so Pet could not live in their home.

Things were great while Pet was little. As she grew, the cat needed more room to run and play. On the first Monday in May, Rhombi found Pet in the sink. "What are you doing, Pet," she laughed and scratched the kitty under her chin.

On the second Tuesday in April, Rhombi found Pet in the cabinet. Pet was a little sulky. She did not want to play. Rhombi started to worry about Pet.

On the third Wednesday in April, Rhombi hopped to stay on her feet when Pet scooted across the room. Pet grumbled a little. On the fourth Thursday in May, Rhombi tripped over Pet three times.

By the fourth week in April, Rhombi and Pet were grumpy about the tiny space. On Sunday of the same week, Rhombi made a decision. For Pet's sake, she needed to give the kitty more space to play.

Rhombi realized that she needed a special place for her new pet. She looked out beyond the pages of her world and asked, "Designers, can you help me build a small house for Pet?"

Davis and Meghan Find a Purrl
Explore perspective in science.

One rainy day, Meghan and Davis were splashing around in mud puddles. They saw a kitten curled up in a cold, sad little ball by the curb. It looked miserable and wet. Meghan tried to pick it up, but the kitten hissed and tried to scratch her arm. Davis pulled off his sweatshirt. They wrapped the angry kitten in Davis's warm sweatshirt. Thank goodness the kitten calmed and began to purr. It fell asleep.

Davis washed and brushed the little black ball of fur. He sat in front of the heater's vent with the kitten on his lap. It was so soft!

The veterinarian talked to Davis and Meghan about healthy food and habits. He made them promise to have the kitten registered and vaccinated. That would keep the kitten from getting diseases later in life. The veterinarian even helped them pick kitten food. They picked a soft food for nutrients and a crunchy food to keep the kitten's teeth healthy.

The kids named their new friend Purrl and got her a bright orange collar. Davis gave the kitten some toys and a bed. He watched Purrl grow taller, and he noticed one day that his pet's eyes had turned green. Purrl got a patch of white fur grew on her chest and white "socks" on her paws. Purrl was growing quickly.

Purrl slept on Davis's pillow every night and begged for breakfast scraps in the morning. As Davis held his purring cat, he whispered, "They all lived happily ever after. Good night, Purrl."

Purrl Finds a Home
Explore perspective in science.

Once upon a time, there was kitten. She was lonely and cold. Rain fell on her fur and dripped in her sad blue eyes. She huddled at the curb trying to get out of the wind.

She saw two giants looking down at her. Cold wet dripped down from giant shiny bodies. Their giant hands reached for her, and she hissed. The kitten's fur stood up in tufts, and her tail stuck straight out behind her body. She hopped, trying to look big and scary. The world got dark and a little bit warmer. Still, the kitten shook and hissed and meowed. She struggled to get free of the dark dry thing, but it was warm and dry. The kitten calmed and began to purr. She was dry. She was warm. She slept.

When she woke, the giants petted her fur. She ate some yummy brown paste and crunchy cookie food. She drank clear, cool water. She curled up by a hot wind and fell asleep again.

The kitten woke to find a thing around her neck. She clawed at the thing. It made jingly noises. She chased it in a circle. The giants laughed. She hissed, but she didn't really mean it…

The kitten washed her fur every day. She grew as tall as the giant's knees. The giants got names. The one with long head fur that smelled like vanilla was Meghan. The one with short head fur that smelled like grass was Davis.

Purrl slept most of the day waiting in the window for sun to touch her paws. That's when the big stinky thing spit Davis out onto the sidewalk every afternoon. Then she and her grassy smell David played until bedtime.

Purrl purred as the two best friends settled in for the night.

Rhombi Lived in a Zoo
Explore habitats and data collection in technology.

Rhombi Lived in a Zoo
Explore habitats and data collection in technology.

There once was a Rhombi who lived in a zoo. She had 26 rescues and life was a hoot. She fed them all dinner with fruits, veggies, protein, dairy, and bread. Rhombi checked on each one before going to bed.

Anna the ape was tucked in her nest.
Bella the boa liked sun warmed rocks best.
Calvin the colt needed plenty of hay.
Dobbie the deer slept someplace new every day.
Ella the Eastern Gorilla slept under the stars.
Fred the ferret curled up in her arms.
Gina Giraffe slept standing up tall.
Hannah the hedgehog curled up very small.
Iggy the inchworm had a leafy bower.
Jasper the jaguar was just waking up at this hour.
Kara the kitten had a basket and blankie.
Her friend **L**atrice Lemur climbed a tree with her hankie.
Mable the mastiff took up a whole bed.
Nonny the newt liked her fishbowl instead.
Odetta Otter burrowed into a holt.
Polly the Pig gets a wallow or away she will bolt.
Qbert the quail must sleep alone,
But **R**ory the rabbit wants friends in his home.
Selena the sloth bunks down on a tree limb.
Tito Terrapin carries his bed around with him.
Una Umbrella Bird has a blanket in her 48 inch cage.
Vixen the fox prefers a crevice with sage.
Wally the wren chooses a nest up high.
XRay the tetra sleeps surrounded by bubbles all the time.
Yuri the yak flops in the dirt on his back.
Zen Zebra curls up in the hay.

There once was a Rhombi who lived in a zoo. She had 26 rescues and life was a hoot. Rhombi makes notes in her file each day. She updates the spreadsheet to show eating habits and play. Who is not sleeping? Who bumped its head? Which animals might want shade or sun instead? Rhombi finished her work, shut down her computer and crawled into bed. "I'll do it all again tomorrow," she happily said.

It All Adds Up
Explore 3D printing in engineering.

P eg watched her dad add layer after layer to the towering party cake. Cake, icing, cake, icing, cake, icing and on and on until the cake was ready to serve. Dad carefully slid the cake off its platform. He removed the little stilts that held the cake still as he iced the layers. Finally, Dad and Poly delivered the finished dessert to the museum for a party.

Dad dropped Peg off at school on the way back to the bakery. Peg ran to her classroom. She was excited about designing for the cool little machines. When Peg arrived at her station in the classroom, she sat down at her wiggly table. She stuffed a piece of folded paper under one of the legs and tried not to get annoyed with the wobbling.

It All Adds Up
Explore 3D printing in engineering.

After fixing the table, Peg found a note. It read, "Innovators look for new ways to solve problems. How will you improve the classroom using our new technology? Sketch. Model. Design. Create!"

Peg's teacher introduced the new printer technology, and explained that these machines were not like a 2D paper and ink printer. 3D Printers used Computer Aided Design (CAD) files to make computer models. 3D printing turns the CAD computer models into real physical things. They melt materials into thin layers on a surface, adding layer on layer until the full object is made. It sounded sort of like making a cake with Dad. He sketched and designed and layered. Leaning on her table, Peg's elbow slipped off as the table wobbled again. "Ugh, this table is so distracting," fumed Peg.

Suddenly, she had an idea. Peg got down on her stomach to take a closer look at the table leg. It was shorter than the other legs, and that made the whole table wobbly. Peg grabbed her logbook and started drawing. She remembered the little stilts her dad used to even out cakes for icing. She measured the length of the other three legs, and drew a sort of stilt for the table. Peg got modeling dough to make a 3D model of her idea. She asked for help in using the scanner and watched as the CAD file was created to turn her idea into a program the printer could read. Her scanned clay model was digitally sliced into thousands of layers in the CAD program. Each of those layers would add up to make her stilt.

Peg held her breath as the first layer of her "Peg Leg" design was melted onto the 3D printer's surface. Slowly, layer by layer, the object was made. At the very end of the day, Peg pulled her "Peg Leg" from the printer's shelf. She broke off the little filaments that held it to the shelf, and hurried to her table. Kneeling, Peg slipped the 3D printed plastic piece under the short table leg. No more wobbly tables! As she headed out to meet her Dad at the car, Peg wondered how she could make her "Peg Leg" even better. Maybe there was a way to change the height? Maybe she could design different shapes for square or round legs. The ideas were endless, and they just kept adding up…

M.C. Escher's World
Explore tessellations in art.

"I don't grow up. In me is the small child of my early days"
– M.C. Escher

Http://www.mcescher.com/about/biography/
Http://euler.slu.edu/Escher/index.php/Regular_Division_of_the_Plane_Drawings
Http://www.csun.edu/~lmp99402/Math_Art/Tessellations/tessellations.html

M.C. Escher's World
Explore tessellations in art.

Escher was a well-known graphic artist. He composed visual material for printing. Escher lived from 1898 to 1972. Like DaVinci and Michelangelo, M.C. Escher was a leftie. He used his left-hand to write and do most activities. Also like these great artists, Escher made a lot of art!

In his lifetime, the graphic artist made 448 lithographs, woodcuts, and wood engravings and over 2,000 drawings and sketches. Escher also illustrated books, designed tapestries, created postage stamps, and painted murals.

After visiting a Moorish castle called Alhambra, Escher became interested in the tiled floors and ceilings. He started drawing special patterns called tiling or tessellations. Escher filled five notebooks with 137 of these tessellation drawings.

Tessellations (or tiling) are created when a shape is repeated over and over again. Each shape fits right up against the next to fill an entire surface. The shapes do not overlap or leave space between them. He described ways to make a tessellation pattern.

You can make pattern pictures like Escher's tessellations using polygons. When you've practiced filling an entire page with geometric shapes, you will be ready to try animals, plants and trees.

ROTATION
turn around a center point

REFLECTION (FLIP)
flip over a line

TRANSLATION
move without rotation or reflection

How Big Is a Guinea Pig
Explore size in mathematics.

How Big Is a Guinea Pig
Explore size in mathematics.

Food, Water, Air, Home.
Guinea Pigs never live alone.

In groups of 5 or 10, wild guinea pigs use ½ inch claws to dig burrows in the ground for sleeping and keeping their tiny pups safe. Sometimes, the little community will use burrows left by other animals or bed down in rocky crevices. The guinea pigs work as a team, screaming to warn each other of predators or other danger.

Like rabbits, guinea pigs have incisor teeth that keep growing through their lives. Eating coarse grasses helps wear down the incisors. Premolars and molars crush and grind food. Also like rabbits, guinea pigs are actually rodents. They are not pigs at all!

Guinea pigs are mammals. Babies, called pups, drink mama sow's milk. Guinea pigs have fur, and they are born live and ready to wiggle. Pups grow quickly. From a few ounces to 2 or 3 pounds in less than a year, a full grown adult might be 10 to 14 inches long. That's about the length of lined paper or legal paper.

Guinea pigs live in their communities eating grass, sleeping through the day in safe burrows, and breathing the clear mountain air of South America from Columbia to Argentina.

Now that you know how guinea pig rodents live in the wild, how would you make a guinea pig happy in a classroom habitat?

HAPPY BIRTHDAY, PET
UNIT 4

HAPPY BIRTHDAY PET
Explore habitats, polygons, and structures.

Once upon a time there was Rhombi. Rhombi had a great playhouse. Pet had a house, too. Both buildings would not be toppled by Autumn winds. The pyramid tops let winter rain and snow slide right off the roof.

Pet had lived with Rhombi for many months. The cat would have her 1st birthday on April 15th. After going to the circus in early April, Pet and Rhombi chose a "Big Top" party theme.

Rhombi was planning Pet's Summer birthday party and wanted to invite all of her barking, squawking, hissing, and mewing friends. Rhombi and Pet printed circus party invitations on primary colored paper.

Pet wanted 7 friends to attend her party. Rhombi invited a dog clown and ordered special cupcakes safe for pets! The third time Pet added a friend to the list, Rhombi realized that she didn't have enough room for a house full of 10 furry, scaly, feathery friends.

Each animal would need four square feet of space to be comfortable. She knew to ask for help. She looked beyond the pages of her world and asked, "Designers, will you help me design and build a covered space for Pet's birthday party?"

That's Bananas
Explore foods in science.

Bananas are actually giant herbs. They do not grow on trees.

A bunch of bananas is called a hand.

The word banana comes from an Arab word "banan" meaning finger.

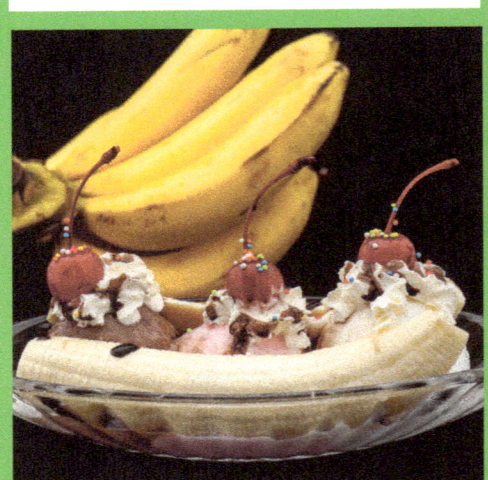

Ecuador is the world's leading exporter of bananas.

An individual banana is called a finger.

That's Bananas
Explore foods in science.

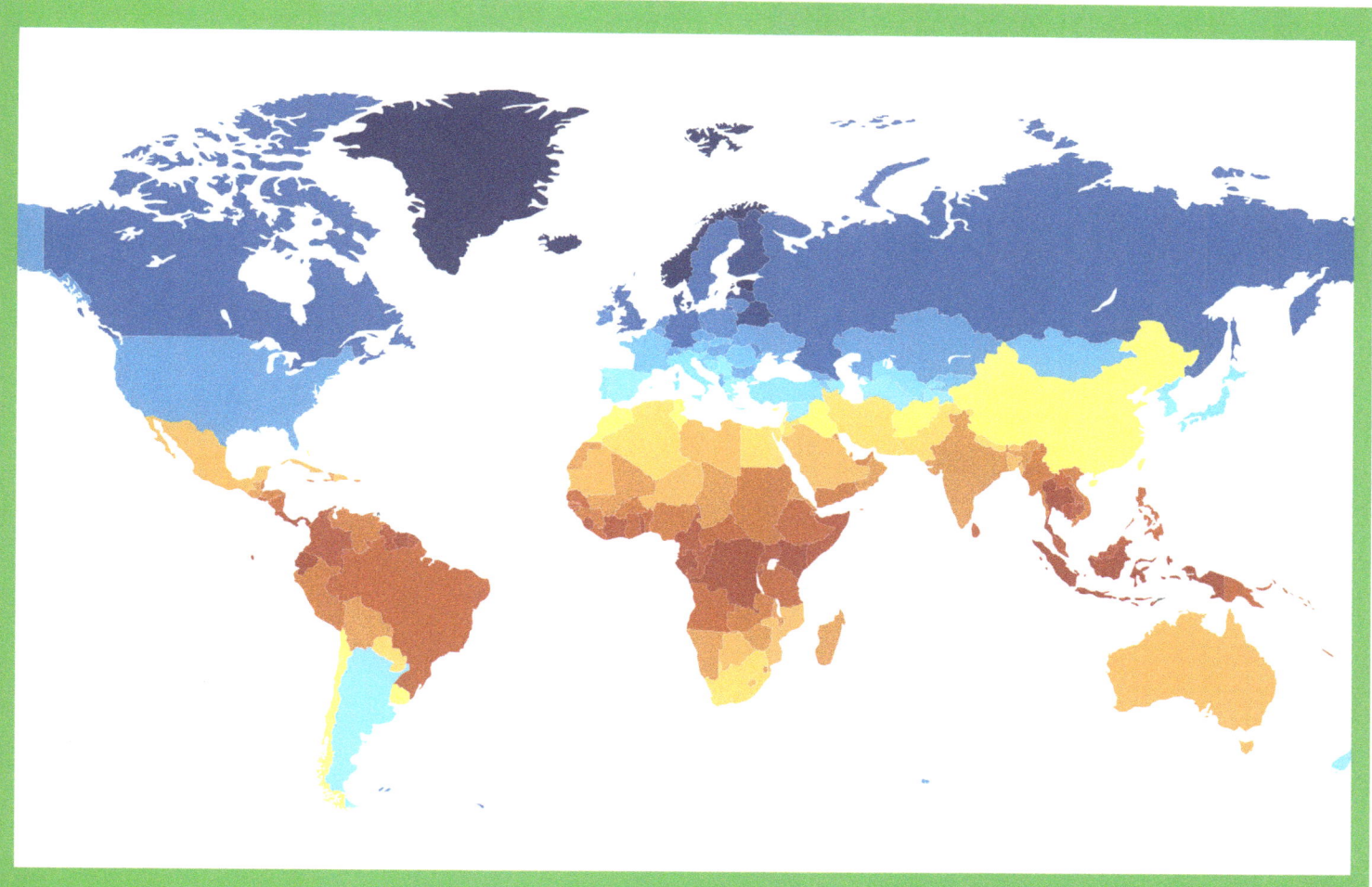

Most of the bananas you buy are grown within 20 degrees on either side of the Earth's equator in Costa Rica, Guatemala, Honduras, and Panama.

http://www.chiquitabananas.com/Banana-Information/banana-fun-facts-and-trivia.aspx

Waxing Colorful
Explore wax in technology.

If you wanted to build a house, what shapes would you use?

Waxing Colorful
Explore wax in technology.

In the hive (or in a wild nest), there are three types of bees:
1. a single female queen bee
2. up to 2,000 male drone bees
3. some 20,000 to 40,000 female worker bees.

The worker bees raise larvae and collect the nectar that will become honey in the hive. When they leave the hive, they collect sugar-rich flower nectar and return.

Honey bees are a little like nurses, a little like architects, and a little like builders. They create wax to build honeycomb in regular hexagons.

The honey bees use a shape that fits together as well as a puzzle. There are no holes or gaps between hexagons in the honeycomb. No wax-making honey or energy is wasted. That also made honey bees sort of like engineers learning the best way to do a job.

The honeycomb "house" idea is working for humans, too. A company in the United Kingdom builds hexagonal houses. Each room of the house fits tightly together to save resources. These houses work for people the way a honeycomb works for bees. The honeycomb shapes take less energy to build, heat and cool.

People have many uses for the beeswax that bees will not use. They make a lot more than any hive needs. Bee farmers harvest the rest for things like candles. Wax from honeybees is also used to make crayons. Most crayons are made of paraffin, but some organic companies are making crayons that contain fewer chemicals. Beeswax is mixed with pigments to make a rainbow of colors. You might be using beeswax candles or crayons in your house!

Strength in Numbers
Explore columns in engineering.

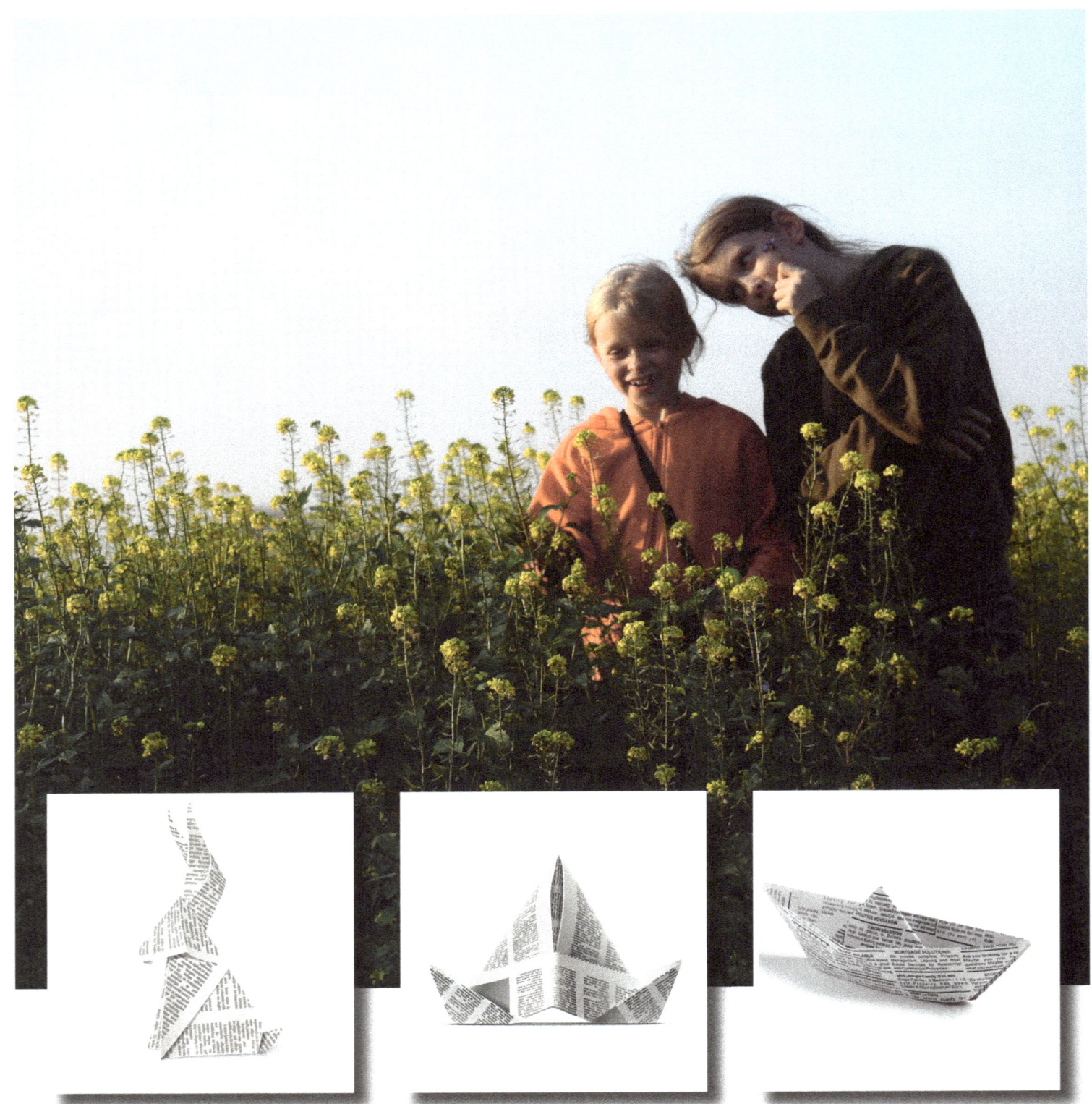

Visit *http://alturl.com/aq5vy* to make a Log Cabin.
Visit *http://www.artistshelpingchildren.org* for folded boats and lots of other great ideas.
Visit *http://www.makingfriends.com* to make hats in under a minute!
Visit *http://www.creativekidsathome.com* to make a dome.

Strength in Numbers
Explore columns in engineering.

Ginny kicked up dust with her sneakers as she headed down the dirt road. She was just wondering if the day could get any hotter when a motion caught her eye. Something was fluttering in the ditch to her left. She walked closer to get a look. She saw that it was most of a newspaper. Ginny did not want to leave litter in the road, so she rolled the paper into a tube and tucked it under her arm. She'd throw it away later.

She was about halfway to her friend's place when the sun broke free of the clouds. It was hot! It was sunny! The glare made her head ache. Ginny remembered the newspaper. She took a page and folded it into a paper hat. Her class had made them for a play last year. She never thought that skill would come in handy. Whew! Her face was cooler in the hat's shade.

Ginny pretended she was a pirate in her buccaneer's hat. She found a stick and heaved it around like a sword. Ginny was doing a pretty good pirate's "arrrr" when she crossed the little bridge over the Spring Road stream. Yesterday's rain had raised the water level, and there was a great current. Ginny stopped to unfold two pages of the newspaper. She folded two small paper boats and set them gently down in the water on the North side of the bridge. She ran to the South side of the bridge to see which boat would arrive first.

She turned back to the road and finished her walk to Meghan's house. Just as she turned into her best friend's driveway, Ginny saw Meghan's father looking for something in his car. Ginny was curious.

"What are you looking for, Mr. Harris?" asked Ginny curiously. He showed her a letter that needed an envelope. Ginny had a handy solution. She used the last page of her newspaper to fold a quick envelope for Mr. Harris. He uses a sticker to close the flap, added a stamp and mailed his letter.

He asked, "Why are you carrying that old newspaper around?"

Pirate Ginny answered, "Because, it's not just a newspaper..."

Square in the Middle
Explore mosaic patterns in art.

Square in the Middle
Explore mosaic patterns in art.

Quad sat glumly on the bench. There was a box on the ground between his feet. When Rhombi sat down, she looked in the box.

"Quad, why are you hauling a box of broken pottery and dishes around the school?"

Quad looked into the box and put his head in his hands. His elbows rested on his knees, and he groaned. "That was my birthday present for Grandad. I dropped it on the way out of the art room, and I don't have enough money to buy another one."

Rhombi pulled the box closer and pushed some of the shards with her pencil. She closed her eyes and began to brainstorm. What could you do with a box and broken plates? Rhombi thought about broken things. Where had she seen broken plates in the last week?

"I have it," cried Rhombi loudly. She giggled and said, "Sorry, that was really loud, but I have an idea." Quad's friend told him about the table her parents had bought for the patio. It had a pattern glued to the top in the shape of the moon and stars. The artist had glued each piece down and filled the spaces with grout.

"You could do the same thing with this box," said Rhombi. "Draw a design, and fill it in with the broken pieces."

Quad sat up and grinned. He liked the idea. Grandad loved cats. Quad could draw a cat on the box. He grabbed the box and headed toward the art room. Turning, he said to Rhombi, "You are a genius!" She smiled and waved him away.

Quad drew a cat. He glued all the broken pieces to the box. The art teacher gave Quad some grout to fill in the cracks between pieces. Mr. Kandinsky said that the grout would be dry in 24 hours. Quad left his masterpiece in the art room window. He was happy with the gift he'd give Grandad, and he knew Grandad would love that he'd made the present all by himself.

Comin' Up a Cloud
Explore materials in mathematics.

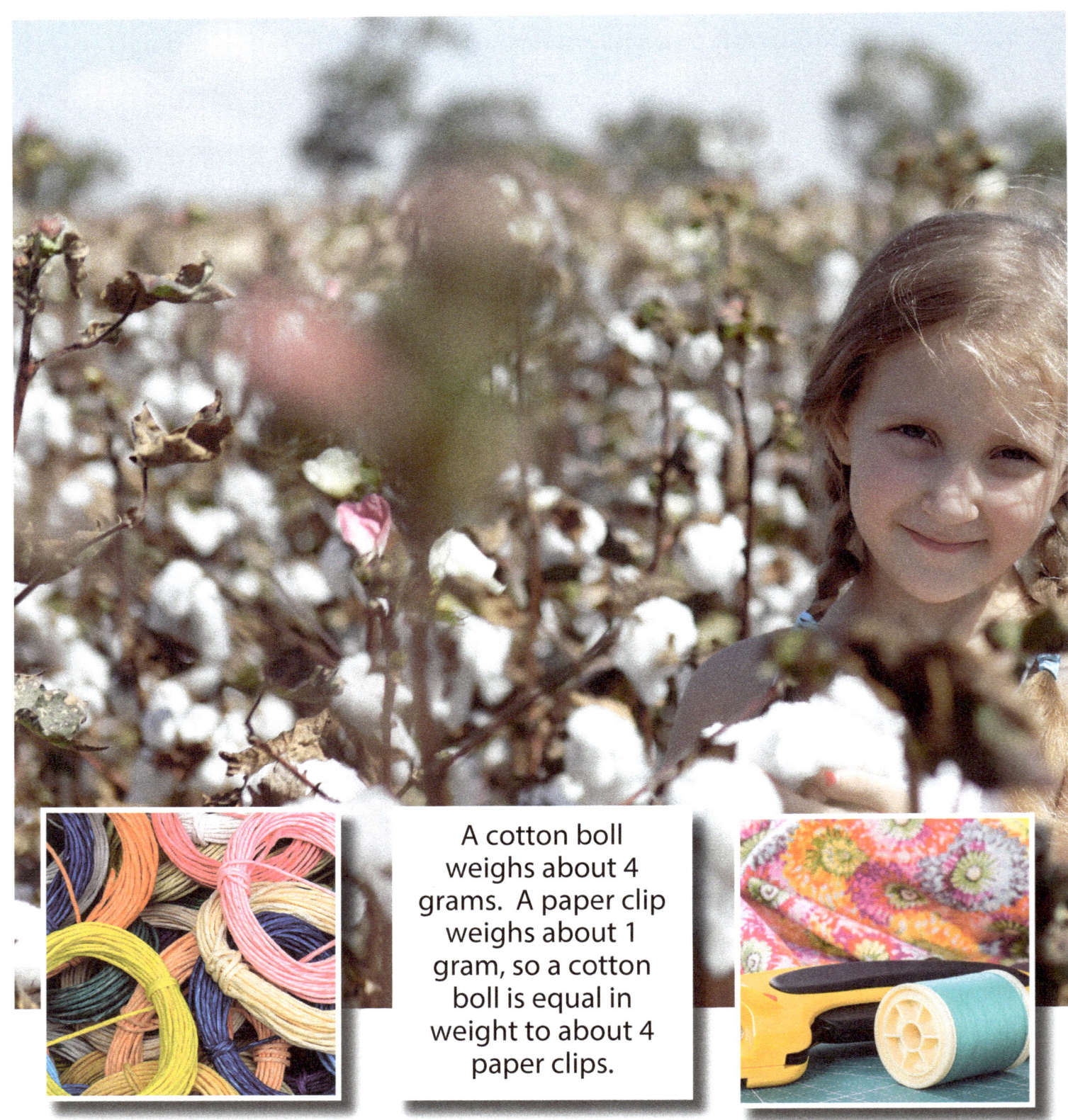

A cotton boll weighs about 4 grams. A paper clip weighs about 1 gram, so a cotton boll is equal in weight to about 4 paper clips.

Comin' Up a Cloud
Explore materials in mathematics.

Ginny carefully wove through the rows of cotton. The Georgia weather was hot and humid. Ginny felt like her clothes were sticky, and the tiny gnats made mad little dashes toward her face and ears. She fanned them away. They didn't bite, but the tiny flies were annoying.

She'd kept an eye on the cotton since planting season in February. The flower buds had opened, and she'd watched as the petals fell to the dusty ground. With petals gone, the rest of the plant had ripened and grown. Each plant had a lot of cotton "bolls" with fluffy fibers.

This week, she and her family would hire local help to harvest the bolls. Then the piles of not-so-fluffy cotton would be "ginned" to get rid of all the seeds. They would send the cotton off to a mill. The mills will mix and clean the cotton. Picking machines will break the piles of cotton into pieces. Dirt has to be shaken out of the cotton fibers.

A carding machine that separates the fibers will comb through the piles. Combing makes sure all the fibers run in the same direction. Ginny's Great-Great Grandma used to do all this by hand, but the machines work with a lot more fiber in a much shorter amount of time. Great-Great Grandma spent weeks doing what a machine finishes in an hour. A long, smooth rope called a sliver forms. The sliver has to be pulled and twisted until it's thin. The thin "roving" is finally ready for a spinning frame. Before it becomes string, all that beautiful roving is twisted and wound onto bobbins.

Ginny wondered where the cotton boll she touched would go. Would it end up on a cargo ship bound for China? Would it land in a pretty gift store? Would she wear it next year as a t-shirt or a pair of jeans? Maybe it would wind her yo-yo or fly a kite. As Ginny bent down to tie her cotton shoelace, a fluffy shadow crossed the field. She glanced at the sky. Weather looked like it might be "comin' up a cloud" as her Grandad always said. Rain might not be good for playing, but it sure did make the gnats go away. Ginny grinned and raced toward the house for dinner.

RHOMBI'S CANDIES EXTENSION

RHOMBI'S CANDIES
Module 2: Extension Unit

Once upon a time, there was Rhombi. Rhombi had a wonderful friend named Pet and a cool cube playhouse with a strong pyramid roof. Pet even had her own house and a party pavilion for friends!

Pet and Rhombi loved 2 dimensional and 3 dimensional shapes of all kinds. Their house was filled with interesting squares, triangles, hexagons, cubes, and pyramids. Rhombi's favorite decoration was a cubist painting in a trapezoid frame.

Rhombi's friend Pet had celebrated her birthday just a few months ago. They had served circus food including cotton candy, lollipops, popcorn, and other pet friendly candy. It had been so popular that Rhombi wanted to recreate the menu at the market.

Rhombi's parents had given permission for Rhombi to open a candy store cart. The cart could be wheeled down to the Saturday farmer's market. When the market ended, the cart could be wheeled right back home.

The farmer's market hosted lots of different foods and crafts. Rhombi's friends Pi and Radius sold fruits and vegetables at the market. She had asked to borrow the cart one weekend and try her business idea.

Pi and Radius had to visit Grandma, so they loaned Rhombi their cart one weekend. She had piled it high with candy and cookies. The cart was open, and all the chocolate melted. Rhombi decided that she needed a new design for her candy cart.

Rhombi looked beyond the pages of her world and asked, "Builders, will you help me design and build a candy store cart with the materials we have?"

FIRST GRADE MODULE 2: 3D SHAPES
VOCABULARY

EARTH AND SPACE	
ASTRONOMER	A PERSON SKILLED IN ASTRONOMY OR WHO MAKES OBSERVATIONS OF CELESTIAL PHENOMENA
APPEARANCE	OUTWARD VIEW BY WHICH AN OBJECT OR THING IS ASSESSED
CHARACTERISTIC	A QUALITY THAT DESCRIBES A LIVING THING OR AN OBJECT
CLOUD	A CLUMP OF TINY WATER DROPS THAT HANGS IN THE AIR
DAY	TIME BETWEEN MORNING AND BEDTIME
MOON	A ROUND BODY THAT ORBITS AROUND A PLANET
NIGHT	TIME BETWEEN BEDTIME AND MORNING
PATTERN	SEQUENCE OF EVENTS
POSITION	AN OBJECT'S LOCATION; WHERE IT IS
SUN	THE STAR CLOSEST STAR TO EARTH AND THE CENTER OF THE SOLAR SYSTEM
PHYSICAL PROPERTIES	
HAND LENSES	A HAND HELD MAGNIFYING GLASS
MASS	THE AMOUNT OF MATTER IN AN OBJECT
MATTER	ANYTHING THAT TAKES UP SPACE AND HAS MASS
NOTEBOOK	A BOOK OF BLANK PAGES FOR NOTES
OBJECT	THING
PATTERN	SHAPES, COLORS, OR LINES PUT TOGETHER IN A CERTAIN WAY
RELATIVE	COMPARING TO
TEXTURE	THE APPEARANCE AND FEEL OF A SURFACE
TOOLS	SOMETHING USEFUL
PHYSICAL PROPERTY	ATTRIBUTES THAT CAN BE OBSERVED AND MEASURED
WATER	
CYCLE	A SERIES OF EVENTS THAT ARE REPEATED OVER AND OVER AGAIN
NATURAL RESOURCE	SOMETHING FOUND IN NATURE THAT PEOPLE CAN USE
SYSTEM	A GROUP OF PARTS THAT WORK TOGETHER AS A UNIT
WATER	A CLEAR LIQUID WITHOUT COLOR, TASTE OR ODOR WHICH ALL LIVING THINGS NEED TO SURVIVE
PATTERN	A REPEATED EVENT
STREAM	A NARROW BODY OF FLOWING WATER
LAKE	A LARGE BODY OF WATER SURROUNDED BY LAND
OCEAN	A LARGE BODY OF SALT WATER

FIRST GRADE MODULE 2: 3D SHAPES
VOCABULARY

MOTION	
ENERGY	ABILITY TO CAUSE MOVEMENT OR CHANGE
MOTION	A CHANGE IN POSITION
POSITION	AN OBJECT'S LOCATION; WHERE SOMETHING IS
CLOSER TO	TO BRING TOGETHER
NEARER TO	TO MOVE TOWARDS A POSITION
FARTHER FROM	TO MOVE AWAY FROM A POSITION
STRAIGHT	NOT BENT OR CURVED
ZIGZAG	MOVING WITH SHARP TURNS
BACK AND FORTH	MOVING ONE WAY; THEN RETURNING
ROUND AND ROUND	MOVING IN A CIRCLE
FAST	TO MOVE QUICKLY
SLOW	TO MOVE AT A LOW SPEED
LIVING THINGS	
BASIC NEED	SOMETHING ALL LIVING THINGS MUST HAVE TO SURVIVE SUCH AS NUTRIENTS, WATER, SUNLIGHT, SPACE, AND AIR
ENVIRONMENT	ALL THE LIVING AND NONLIVING THINGS IN AN AREA
LIFE CYCLE (PLANT)	STAGES IN THE DEVELOPMENT OF A PLANT
LIVING	SOMETHING THAT HAS BASIC NEEDS AND PRODUCES OFFSPRING
NONLIVING	NOT ALIVE, DOES NOT HAVE BASIC NEEDS OR PRODUCE OFFSPRING
OFFSPRING	NEW ORGANISMS THAT COME FROM PARENT ORGANISMS
ORGANISM	A LIVING THING
CAREERS	
CONTRACTOR	A GENERAL CONTRACTOR IS RESPONSIBLE FOR THE DAY-TO-DAY OPERATIONS OF A CONSTRUCTION SITE. CONTRACTORS MANAGE SALES PEOPLE, BUILDERS, PLUMBERS, ELECTRICIANS, AND ALL THE OTHER TRADES REQUIRED TO COMPLETE A PROJECT. THE CONTRACTOR ALSO COMMUNICATES INFORMATION TO ALL THE PEOPLE INVOLVED IN THE PROJECT.
DESIGNER	INDUSTRIAL DESIGNERS DEVELOP CONCEPTS AND SPECIFICATIONS THROUGH COLLECTION, ANALYSIS AND SYNTHESIS OF DATA GUIDED BY THE SPECIAL REQUIREMENTS OF THE CLIENT OR MANUFACTURER. DESIGNERS MAKE CLEAR AND CONCISE RECOMMENDATIONS THROUGH DRAWINGS, VERBAL DESCRIPTIONS, AND MATH MODELS.

Log onto the website to access additional lessons, activities, word wall printables, printable nets, videos, downloadable files, links, and other resources.

www.ten80elementary.com

Please contact us for more information about curriculum, programs, festivals, and professional development.

info@ten80elementary.com